THE SAGUARO CACTUS

Gates Pass, Tucson Mountains, February 22, 2019. Photo by David Yetman.

The Southwest Center Series

JOSEPH C. WILDER, EDITOR

THE

SAGUARO CACTUS

A NATURAL HISTORY

DAVID YETMAN,
ALBERTO BÚRQUEZ,
KEVIN HULTINE, AND
MICHAEL SANDERSON
WITH FRANK S. CROSSWHITE

THE UNIVERSITY OF
ARIZONA PRESS
TUCSON

The University of Arizona Press
www.uapress.arizona.edu

ISBN-13: 978-0-8165-4004-4 (paper)

Cover design by Leigh McDonald
Cover photo: *Picture Rocks Sunset* by Jesse Jackson / JesseJ.media
Designed and typeset by Leigh McDonald in Fournier and Minion Pro, 10.25/14

Library of Congress Control Number: 2019952316

Printed in the United States of America
♾ This paper meets the requirements of ANSI/NISO Z39.48–1992 (Permanence of Paper).

We dedicate this volume to the memory of our mentor and friend Raymond M. Turner, whose studies on the saguaro cactus were an inspiration and an unending source of information. His willingness to engage students, his magnanimous personality, and his profound understanding of the Sonoran Desert provided an unmatched influence in our thinking and in our writing.

FIGURE 1.1 A rather old saguaro near Mammoth, Arizona, in the San Pedro Valley, a location of many old and large plants. The buds and flowers indicate that the photo was taken in late spring, probably late April or early May. The branch at the center has three cavities carved by Gila woodpeckers or gilded flickers. The highest of the three appears to be the most recent and may presage the breaking of the branch above that point. Other branches support hollows as well. The numerous branches and cavities indicate it is of great age. This individual is probably in the last decade or two of its life. Photo by David Yetman.

THE CANE MAKER

He made a cane from a saguaro rib so his wife
would walk the world with him, and so the rib,
its ancient flesh sloughed away, would rise
straight and steady beneath her delicate hand.

The rib walked beside her and remembered
the rain, the cactus wren, the white flowers,
the red fruits—their seeds planning giants
to grace the desert skyline.

That night the rib dreamed itself whole again.
It appeared at dawn rooted through the floor,
busting through the roof into familiar blue—
a forty-foot column of thorned water.

Everyday oak would have to do—
store-bought, rubber tipped, merely useful.
A workable rib would come eventually.
The cane maker shelved his tools and waited.

—NED MACKEY

CONTENTS

.

ACKNOWLEDGMENTS

I wish to acknowledge the special assistance provided by Janelle Weakly
of the Arizona State Museum and Debra Colodner and Craig Ivanyi of
the Arizona-Sonora Desert Museum.

—DY

I wish to note the special assistance from the following individuals:
Dario Copetti, Marty Wojciechowski, and Michelle McMahon.

—MS

I wish to acknowledge the special assistance from the following individuals and institutions: Angelina Martínez-Yrízar, Enriquena Bustamante,
Dirección General de Apoyos al Personal Académico-UNAM, and the
Consortium for Arizona-Mexico Arid Environments (CAZMEX).

—AB

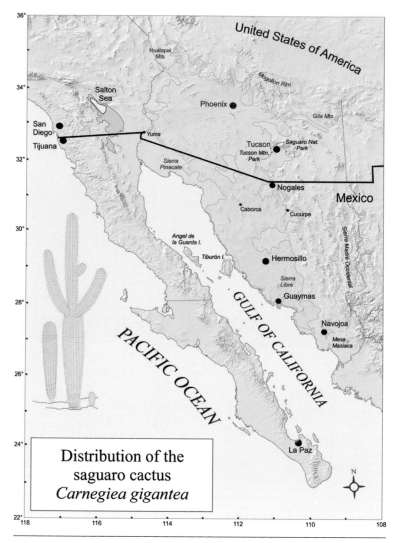

MAP 1.1 The pale green color represents the distribution of the saguaro cactus. The region south of Guaymas is beyond the generally recognized limits of the Sonoran Desert, with a notable disjunct population on Mesa Masiaca. Note also the slight extensions into the state of California and the absence from Baja California. Map by Alberto Búrquez.

THE SAGUARO CACTUS

One

A SAGUARO PRIMER

Carnegiea gigantea in History

DAVID YETMAN

*As we approached the foot of the botanical mountain, I framed up
a foolish question. I was about to say, "Why are your fence posts
so tall, and so irregular?" But for once I wisely held my peace;
and presently it was clear that all those seeming tall straight posts
running up the mountain on the southern sky-line were giant cacti,
without sidearms. They stood all over the plain, and climbed up all
sides of the mountain.[1]*

—WILLIAM HORNADAY, UPON MAKING HIS
FIRST VISIT TO THE DESERT LABORATORY
ON TUMAMOC HILL IN 1907

*More amazing perhaps than any aspect of its biology is Man's
emotional involvement with the saguaro—the saguaro is a "hero"
among plants. He has endowed it with human attributes and be-
stowed upon it affection and concern for its "problems."*

—SCOTTY STEENBERGH AND CHARLES LOWE,
ECOLOGY OF THE SAGUARO II

The saguaro is one of the world's most studied wild plant species, probably because of its charisma and ready accessibility, but also because, as a large cactus, it is decidedly different from plants more familiar to researchers. Virtually every aspect of its life and its place in cactus evolution has been minutely examined. The essays in this book bear witness to the ongoing fascination researchers find in the great cactus, as well as the plant's unusual characteristics. There is simply no other plant like it in the United States, so instantly identifiable, so predictably located, and possessor of such a variety of distinct characteristics. It also has a national park dedicated to it, joining only the Joshua tree, the coastal redwood, and the sequoia in that honor. The vast literature concerning the saguaro is testimony to its prominence as a symbol, to the perceptions it inspires, its role in human society, and its role in desert ecology.

The saguaro, with its great size and characteristic shape—its arms stretching heavenward, its silhouette often resembling a human—has become the emblem of the Sonoran Desert of southwestern Arizona. This is rightly so, for it is by far the largest and tallest cactus in the United States and our tallest desert plant as well. In this volume, we present a summary of current information about this, the desert's most noteworthy plant.

Saguaros occasionally reach 12 m (40 feet) in height, and individuals over 15 m (50 feet) tall appear from time to time. The record height is 23 m (78 feet), a well-known plant of a single stalk growing near Cave Creek, Arizona, which was toppled by winds in 1986. Photos of that plant are elusive, but it was clearly a very tall cactus, perhaps the tallest of any cactus ever recorded. While other cactus species may produce individuals taller than the average saguaro, none has been documented of that stupendous height. In 1907 William Hornaday reported a saguaro between 55 and 60 feet in height.[2] He was leader of a 1907 scientific expedition to Pinacate Volcanic Range in Mexico near the border with southwestern Arizona and was in the company of distinguished researchers. The saguaro's sole competitor for tallness in the deserts of the United States is the Joshua tree (*Yucca brevifolia*), a native of the Mohave Desert, a yucca that only rarely reaches 9 m (30 feet) in height.

Saguaros are among the tallest cacti in terms of average height. They are also among those with the greatest mass. *Neobuxbaumia mezcalaensis* of southern Mexico, a single-stalked columnar cactus and distant relative of the saguaro, probably reaches greater average height, with individuals reaching in excess of 18 m (60 feet). Other columnar plants include *Pachycereus weberi* and *Mitrocereus fulviceps* of southern Mexico and *Pachycereus pringlei*, the *cardón sahueso* of the Sonoran Desert in Baja California and the coastal regions and islands of central Sonora. *Pachycereus pringlei* and the truly massive *P. weberi* routinely exceed the mass of the saguaro.[3] While columnar cacti are widespread in South America, none reaches the height or mass of the larger saguaros.

The most famous incident involving cacti of any kind occurred in 1982. The episode featured a saguaro growing near Phoenix, Arizona, and an unfortunate drunk named David Grundman, a hapless chap. Grundman, having imbibed an excess of strong drink, decided to knock over a saguaro with his jeep. He failed, succeeding only in damaging his vehicle. In a fit of rage at the unobliging saguaro, he fired both barrels of a shotgun at its base. The blast weakened the trunk, and the great plant toppled, crushing Grundman beneath. Few observers shed tears over the vandal's demise. A published ballad commemorates his folly.

Part of the allure of saguaros, attributable to their towering height, is that they are a truly tree-size cactus, the only such plant in the United States.[4] They are unmistakable, truly incapable of being confused with any other plant growing in this country, and they lend themselves to artistic depiction like no other plant. They appear dramatically on the landscape as the visitor enters the Sonoran Desert of the United States and disappear just as suddenly beyond that desert's boundaries, except in the extreme south.

People indigenous to the Sonoran Desert region have been intimately familiar with saguaros for millennia. For many of them, especially the Tohono O'odham, it was their most important plant, providing food, drink, lumber, tools, and shade. The O'odham term for saguaro is *haashañ*. They refer to the Slate Mountains, toward the northern edge of the O'odham Reservation as *Haashañikam*, or "Place where saguaros grow in abundance."[5] Seris, people of the coast of Sonora who are geographically close to the O'odham but linguistically and culturally distant, call the saguaro *mojepe*. They rely extensively on the plants for fruits and building materials. Mayos and Yaquis, whose lands lie at the southern edge of the Sonoran Desert and the saguaros' range, call it *saguo*, a word from which the common name seems to be derived. Sonoran historian and linguist Horacio Sobarzo notes that traditional Sonorans formerly referred to a tall man of elegant posture as a *saguarón*, or "big saguaro."[6]

Spaniards arrived in the Sonoran Desert on foot and on horseback from southern Mexico, where they had already become familiar with Mexico's variety of cacti, including an impressive array of columnar cacti and had noted the important role that cacti played in pre-Columbian cultures of New Spain. They labeled any columnar cactus a *cardón*, or big thistle, a name still applied to a variety of columnar cacti in Latin America. Others they described as *órganos*, apparently for their resemblance to organ pipes, others still as *candelabro*, impressed by their resemblance to candelabras.

From the sixteenth century onward, Spanish and, later, Mexican botanists collected and classified cacti with scrupulous precision. Their work was rendered less onerous by the expertise of native botanists, who long ago had catalogued cacti according to their appearance and their usefulness. The Spaniard Francisco Hernández, who studied the medicinal uses of plants in New Spain and published his findings in 1651, chronicled this knowledge. From the first expeditions to the Americas, Spaniards had been enchanted with cacti and transplanted large numbers of plants to gardens in Europe, where European botanists became acquainted with them. Postconquest documents from Mexico are replete with references to cacti, often referred to as *nochtli*, the Náhuatl (Aztec) generic term for them.

In the mid-eighteenth century, the Swedish botanist Carl Linnaeus proposed a new order of plant and animal taxonomy that would gain worldwide acceptance, that of binomial nomenclature. His system provided a single standard of classification that crossed linguistic and ideological boundaries and imposed order on the world of botanists. Among the plant families he recognized was the Cactaceae (*Cactos*). And thus, when a cohort of European-trained taxonomists and plant explorers visited the landscapes of New Spain and catalogued its botanical wonders, their nomenclature was published in a form still useful to botanists. Included in their numbers was the eminent Mexican statesman Melchor Ocampo, who published a classification of cacti in 1844.[7]

In the 1780s King Charles III of Spain commissioned a multidisciplinary, but primarily botanical, expedition to carry out a comprehensive inventory of the riches of New Spain. The result was a sixteen-year project carried out between 1787 and 1803 that extended from Vancouver Island to San Francisco (still a part of the Spanish Empire) to Nicaragua. The leaders of the expeditions were the Spaniard Martín Sessé and the Mexican José Mariano Mociño, noted botanists whose species descriptions survive to this day.[8] They readily imposed Linnaeus's system onto their extraordinary collection of plants. The result was a folio of incomparably fine drawings of plants. Many species' descriptions still bear the authors' designation *Sessé & Moc.* Their expedition (or expeditions, for there were multiple forays led by different leaders) was the most ambitious and encompassing ever launched in the Americas.

It is unclear which portions, if any, of northwest New Spain were included in the Sessé and Mociño expeditions, and so we have no record of the saguaro being catalogued. Members traveled as far north as Vancouver Island, but perhaps more for political and geographical reasons than to collect plants, and did not appear to venture far inland. A separate journey may have reached the southern limits of the saguaro's range, but no plant vouchers or records have been recorded from Sonora.

If members had traveled inland from San Francisco to Mexico City, they would almost certainly have penetrated the Sonoran Desert and traveled through forests of saguaros. Captain Juan Bautista de Anza had traveled in the opposite direction, from central Sonora to the San Francisco Bay area only a decade earlier (1775), ultimately founding the city. Anza's expedition departed from the hamlet of San Miguel de Horcasitas, Sonora, in the heart of saguaro country, and he and his expedition of about 250 people would have been intimately familiar with saguaros and their uses. A Franciscan priest, Pedro Font compiled a voluminous diary of the expedition, but paid little attention to natural history. He did record a myth related by the River Pimas (Akimel O'odham), who lived near the confluence of the Santa Cruz and Gila Rivers. He reports that the Indians spoke of a powerful figure named the Drinker, who arrived in their country.

At one time the Drinker became angry with the people and killed a great many of them and changed them into saguaros and this is why there are so many saguaros in the country. (The saguaro is a green, watery trunk of considerable height and evenly round and straight from its foot to its top, with rows of thick thorns all the way up, and usually having two or three branches formed in the same fashion, that look like arms.)[9]

The Anza expedition passed through the Akimel O'odham and Maricopa lands in late October 1775, so they did not witness the flowering or fruiting periods, which probably would have attracted the priest's attention, if only to disparage the natives' use of saguaro wine. Ignaz Pfefferkorn, Jesuit missionary in the 1760s first at Tubutama on the Río Altar, then at Cucurpe, on the Río San Miguel, both in northern Sonora in areas thick with saguaros, wrote substantially on the geography of Sonora, including its plants, but had little to say about the saguaro. He pronounced it to be similar in size to the organ pipe cactus, a clear mistake on his part.

Earlier records of saguaros are mostly sketchy. By the time Spaniards arrived in Mexico's northwest for good, beginning in the 1620s, they had endured hardship, heat, and many unreceptive Indians. Furthermore, they had already encountered a large variety of columnar cacti in their journeys from the landing at Veracruz, so the spiny giants were no novelty. Apart from a couple of observant Jesuit missionaries, few of the early immigrants recorded more than passing descriptions of saguaros in their notes. Sonora lay roughly 2,000 km from Mexico City, all of it on horseback or foot, and the region was (from the standpoint of Spanish or, later, Mexican interests) constantly plagued by Indian rebellions, making settled life difficult and unattractive to plant collectors, who would have to have been a hardy bunch. The remarkable plant diversity of central and southern Mexico was sufficient to absorb the work of a phalanx of botanists. The northwest was put off till later.

For explorers and agents arriving from the United States, however, saguaros were the first columnar cacti they encountered, and their impressions were noteworthy. It was not until an event of international importance that we have the earliest sustained descriptions of saguaros: reports from the Boundary Survey Commission, charged with fixing the boundary between Mexico and the United States.

In the war against Mexico, beginning in 1843, U.S. military domination forced Mexico into the Treaty of Guadalupe Hidalgo of 1848, in which Mexico ceded an enormous chunk of what is now the Southwest to the United States for about $15 million. The new U.S. territory included all of what are now Arizona and New Mexico and much of California, plus parts of Colorado, Nevada, Utah, and Wyoming. The annexed lands also included a significant portion of saguaro habitat—all those plants located north of the Gila River. John Russell Bartlett, a renowned man of letters, headed the early Boundary Survey Commission, a party dispatched by the U.S. government to determine just what the United States had got for its money and military success and to fix the border. From Bartlett's field notes, later published privately as his *Personal Narrative*, appeared the first widely circulated English language popular descriptions of the saguaro. While Bartlett was not specifically instructed to submit geographical descriptions to the U.S. government, he was a keen observer of nature, and his diary and drawings brought to public attention the peculiar and wondrous plant known as the saguaro and its fruits, which he referred to as *petahayas*.

General W. H. Emory, whose notes describing the saguaro were among the first to be circulated among botanists, succeeded Bartlett as director of the Boundary Survey Commission, perhaps in 1853. Emory had a passion for collecting specimens for scientific study, and his enthusiasm infected members of the survey team. Under his direction, they also gathered thousands of plants and pressed and shipped them for description or identification to botanists stationed at leading institutional herbaria in the eastern part of the United States. Other officials of the survey also contributed botanical collections from the Sonoran Desert region, many species of which still bear their names, such as Bigelow, Parry, Schott, Thurber, and Wright. For the cactophile (a person with unusual fondness for cacti), the most useful publication to emerge from the survey is *Cactaceae of the Boundary*, published in 1858 by Dr. George Engelmann, after whom

FIGURE 1.2 Drawing from John Russell Bartlett, who incorrectly labeled saguaro fruits *petahayas*. Since the O'odham revere plants and the fruits, some historians question whether the Native Americans depicted are O'odham and not adjacent groups, such as Maricopas.

shows, and assumes the shape of a small carrot, almost as large as the stem itself; in old specimens the root is very much larger than the whole stem and branches together.

Subgen. 3. LEPIDOCEREUS.*

22. C. GIGANTEUS, E. in Emory's Rep. 1848: erectus, elatus, cylindricus, versus basin apicemque sensim attenuatus, simplex seu parce ramosus, candelabriformis ; ramis paucis erectis caule brevioribus ; vertice applanato tomentoso; costis infra sub-13 sursum 18–21 rectis, vetustis (versus caulis basin) obtusis obtusissimisque, sursum e basi lata acutatis acie obtusatis subrepandis ; sinubus ad basin caulis latissimis versus apicem profundis acutis angustioribus angustissimisque ; areolis prominentibus ovato-orbiculatis, junioribus albido-tomentosis ; aculeis rectis basi valde bulbosis tenuiter sulcatis et subangulatis albidis seu stramineis demum cinereis ; radialibus 12–16, imo summisque brevioribus, lateralibus (præcipue inferioribus) longioribus robustioribus, subinde aculeis adventitiis paucis setaceis summo areolæ margini adjectis ; aculeis centralibus 6 robustis albidis basi nigris apice rubellis demum totis cinereis, 4 inferioribus cruciatis quorum infimus longissimus validissimus deflexus, 2 superioribus brevioribus lateralibus sursum divergentibus ; floribus versus apicem caulis ramorumque aggregatis ; ovarii ovati sepalis 30–40 squamiformibus triangulatis acutis ad axillam albido, seu fulvo-villosam aculeolum unum alterumve nigricantem deciduum gerentibus ; sepalis tubi ampliati breviusculi 30–40 orbiculato-subdeltoideis mucronatis, inferioribus in axilla lanigeris, superioribus nudis ; sepalis intimis 10–15 spathulatis obtusis carnosis (pallide viridibus albescentibus) ; petalis sub-25 obovato-spathulatis obtusis integris crispatis coriaceo-carnosis crassis (ochroleucis seu albidis) ; staminibus numerosissimis superiori tubi parti adnatis, inferiore nudo ; stylo stamina paullo superante ; stigmatibus 14–18 filiformibus fasciculatis ; bacca obovata seu sæpe pyriformi squamis triangulatis carnosis parvis ad axillam lanatis munita, floris rudimentis deciduis ; pericarpio duriusculo coriaceo demum valvis 3–4 irregularibus patulis reflexisve dehiscente ; seminibus numerosissimis in pulpa saccharina coccinea nidulantibus oblique obovatis lævibus lucidis exalbuminosis; hilo oblongo basilari ; cotyledonibus foliaceis incumbentibus hamatis. (Tab. LXI, LXII, et tab. front.)

In rocky valleys and on mountain sides, often in mere crevices of rocks, from the valley of Williams' river, lat. 35°, *Bigelow*, to Sonora, lat. 30°, *Thurber, Schott ;* and from the middle Gila, *Emory*, down to near its mouth, *Parry.* I cannot find that it has been collected west of the Colorado, though it is probably also an inhabitant of the Californian peninsula : fl. May to July ; fruit ripe in July and August. The *Suwarrow* or *Saguaro* of the natives.—Young plants, as *Dr. Bigelow* observed, are almost always found under the protecting shade of some shrub, especially of Frémont's "green-barked Acacia" (*Cercidium Floridanum*) so characteristic of the barren wilderness ; and not rarely the dead stems of this plant are found near the older *Cerei.* Young plants retain their globose shape for several years ; a specimen in my possession, 5 or 6 inches high, is supposed to be between 8 and 10 years old. It flowers at the height of 10 or 12 feet, but grows up to 4 or 5 times that height ; stems have been measured 46 feet high, and higher ones are stated to occur, so that the first statement of *Col. Emory* is not at all improbable, viz : that this plant sometimes has been found 50–60 feet high. The stem is thickest in or a little above the

---

* This subgenus is proposed for the two tall western species with uniform spines, short flowers, ovary and tube with numerous scale-like imbricated sepals, fleshy petals, pale stigmata, smooth seeds, and hooked embryo. Probably *C. Chilensis*, and perhaps other species from the Pacific slope of the continent, will find their place here. A drawing of *C. Chilensis*, among the papers of the United States Exploring Expedition, represents a flower almost identical with that of *C. Thurberi.*

FIGURE 1.3 Engelmann's published scientific description of the saguaro from 1858.

FIGURE 1.4 Engelmann's drawing of a saguaro. It greatly exaggerates the plants' size, perhaps deliberately so.

the Engelmann spruce is named. A prominent botanist trained as a medical doctor who could also serve as physician for the survey team, Engelmann's taxonomy of the saguaro is sophisticated: he wrote the original species description, which to this day bears his name. His original description of the plant (*Cereus giganteus*), first published in 1848,[10] is remarkably detailed and his ecological observations accurate and informative. The drawing of the plant included in his publication, however, is a tad romantic and vastly exaggerates the saguaro's size.

The boundaries resulting from the Treaty of Guadalupe Hidalgo did not include the portion of desert territory to the south that surveyors considered the best route for constructing a railroad connecting the eastern and southwestern United States. As a result, negotiations between the two nations began almost immediately after the signing of the Treaty of Guadalupe Hidalgo for the United States to acquire much of what is now southern New Mexico and southern

FIGURE 1.5 Saguaros and desert vegetation, Organ Pipe National Monument, an area included in the Gadsden Purchase. Photo by Alberto Búrquez.

Arizona—land that afforded the least mountainous and safest route for a railroad. While many Mexicans and U.S. citizens protested any further annexation of Mexican territory by the ambitious nation to the north, the Mexican government needed money and consequently agreed to sell the parcel to the United States. The transaction was completed in 1853 in a treaty known as the Gadsden Purchase in the United States and *Tratado de la Mesilla* in Mexico. The acquisition fixed the current border of Arizona and New Mexico with Mexico. Establishing its precise location fell to the existing boundary commission, which simply added the Gadsden Purchase to its survey package. And thus, the densest saguaro forests came to reside within the boundaries of the United States.

The construction of the Southern Pacific railroad through the Sonoran Desert region, as envisioned by the promoters of the Gadsden Purchase, was completed in 1880. Almost overnight, passenger service made arrival in the desert far easier for the armed forces, land speculators, adventurers, and sufferers of asthma and tuberculosis, long before the advent of improved roads. It also opened a vast area of research for students of natural history. Scholars and scientists could board the railroad on the East Coast and arrive in southern Arizona four days later. In 1903, scientists associated with the Carnegie Institution founded the Desert Laboratory in Tucson, Arizona, an important stop along the Southern Pacific route. The laboratory was constructed on a high shoulder of a volcanic peak called Tumamoc Hill, which overlooks the Santa Cruz River (in reality, a small stream at the time, now a dry wash except for occasional ephemeral runoff) and lies only a couple of miles from the city's center. The founding purpose of the Desert Laboratory was to promote research into the relationships among plants and animals and the desert in which they lived. It seems certain that the driving force surrounding the creation of the laboratory was scientists' fascination with the saguaro, which abounds on Tumamoc Hill, and it is around this time that scholarly descriptions of the ecology of the saguaro cactus began appearing in learned journals.

No scientist could match the importance of Forrest Shreve in describing the ecology of deserts, especially the Sonoran Desert. He can be safely labeled the founder of Sonoran Desert ecology. It was he who first proposed specific boundaries to the Sonoran Desert, limits he suggested based on his own expeditions and fieldwork. Shreve moved to Tucson in 1908 to join the Desert Laboratory and used it as a base for field studies that took him to nearly every landscape in the desert Southwest, northwest Mexico, and Baja California. His first publications on the saguaro and desert ecology appeared in 1910, and he published almost continuously until his death in 1950. The first of his two-volume *Vegetation and Flora of the Sonoran Desert* was published posthumously in 1951. Volume 2 was completed by his collaborator Ira Wiggins and published in 1964. The work contains the results of Shreve's decades-long studies of the Sonoran Desert, its plants, and its vegetation, along with the flora of the region compiled by Wiggins. While Shreve's writing is somewhat dry and his pronouncements clinical, the saguaro emerges as the most important plant in his vast studies.[11]

Perhaps the most important study of cacti emerged from the work of Nathanial L. Britton and Joseph N. Rose. Between 1919 and 1923, they published a magisterial four-volume, lavishly illustrated work, *The Cactaceae*, funded in part by the Carnegie Institution. The Desert Laboratory was influential in the volumes' scope. Many of Britton and Rose's descriptions remain intact, in spite of seemingly unending revisions, lumpings, and splittings by subsequent taxonomists.[12]

It is also important to acknowledge the profound expansion of knowledge of Mexican cacti resulting from the work of Mexican cactologist Helia Bravo-Hollis. In 1978, Mexico's National Autonomous University published the first volume of her three-volume work *Las Cactáceas de México*, which she had begun in the 1930s. The third volume was published in 1991 with the assistance of the eminent Mexican cactologist Hernando Sánchez-Mejorada and remains an invaluable resource for students of cacti and of the distribution of the family in Mexico, though it has yet to be translated into English.[13]

Since Shreve's publications, numerous scientists have published studies of the ecology, biochemistry, phenology, and biogeography of the saguaro. The plant seems an irresistible object of study from a multitude of perspectives, as this volume demonstrates.

The history of political control of the saguaros' habitat does not reflect well on the foreign policy of the United States (a substantial percentage of U.S. citizens opposed both the Mexican War and the acquisition of the Gadsden Purchase), but it has been fortuitous for saguaros. Local, state, and national institutions now protect most of the finest saguaro habitat in the United States, while none of the most prolific populations in Mexico appears to have any protection apart from that afforded by private landowners.

In the ensuing chapters, we present a comprehensive discussion of the saguaro. In chapter 2, Alberto Búrquez and I describe the cactus family, a large assemblage of plants native to the Americas, and locate the saguaro's place in that family. In chapter 3, we discuss the fit of the saguaro in the realm of plants and animals, the ecological role of the saguaro. In chapter 4, Kevin Hultine describes in detail how the saguaro is put together and how the great plants manage to flourish and grow to giant size in the heat and drought of the Sonoran Desert. In chapter 5, Michael Sanderson takes us into the realm of saguaro genetics, demonstrating how scientists have determined how the saguaro's genome compares and relates to other cacti and how we go about finding out about the plants' previously hidden past. Finally, we present a reprint of a 1980 article that has proved the classic description of the ethnobotany of the saguaro among the Tohono O'odham, the people for whom the saguaro meant the difference between life and death.

Two

CACTACEAE

The Cactus Family, Columnar Cacti, and the Saguaro

DAVID YETMAN AND ALBERTO BÚRQUEZ

Cacti are natives of the Americas, that is, the Western Hemisphere, which includes the Caribbean islands. It seems odd that such a large and varied family—somewhere between 1,400 and 1,800 species—should be so geographically confined. Only three species naturally occur beyond the Western Hemisphere. Those are peculiar sorts of plants that have managed to cross the Atlantic Ocean from South America to Africa and some points east. But they are subdued cacti, arboreal (tree-dwelling, epiphytic) species with especially sticky seeds. At some point seeds must have hitched a ride on a bird that was devouring the cactus fruit in South America and made it across the Atlantic to the Eastern Hemisphere. Or perhaps humans unwittingly transported the seeds during the slave trade period. Or maybe ocean currents bore them across the Atlantic. It cannot have been long ago, for divergence of those three into multiple species has yet to occur. Evolution of cacti in Africa is just beginning to crank up.[1]

The evolution of cacti can be traced to the cooling climates of the mid-Tertiary period. The Late Cenozoic ice age began 34 million years ago (MYA) at the boundary between the Eocene (56–34 MYA) and the Oligocene (34–23 MYA) epochs and continued into the Miocene (23–5 MYA). This global trend continued from a warmer Oligocene to a cooler Pliocene (5–2.5 MYA) and resulted in enormous consequences for the evolution of organisms. The newly emerging worldwide climates saw the establishment of new taxonomic groups, new ecological biomes, and new combinations of functional traits, all of them related to evolutionary novelties, strategies to deal with a harsher environment. We live in times of dramatic global warming and of vanished and vanishing glaciers, but our earthly climates are cool and varied when compared with the Eocene.

The clash of continents through tectonic movement is directly implicated in these alterations in Earth's climate. By the Early Miocene, the rise of the Himalayas, which resulted when the Indian subcontinent crashed into the Asian landmass, was blocking the previous circulation of air currents. This dramatic alteration directly affected the Asian monsoon and indirectly altered the global climate, leading the planet toward a drier and cooler condition. And the new regimes worked to our advantage, for while plants had long since scurried to fill new niches, apes also diversified, and by the end of this epoch, the ancestors of humans and chimpanzees—our closest living relatives—split apart to follow different trajectories. Meanwhile, forest cover diminished on a global scale, while drylands acreage increased, frequently merging into ecosystems such as grasslands and deserts. In the sea, kelp forests developed, nurturing and harboring a host of new species in a new kind of sea biome. And this planet-wide drying out led to the evolution of plant and animal families adapted to the greater aridity and/or cooler temperatures. For example, horses rapidly evolved to cope with the new steppe environment of North America, enormous birds weighting up to 100 kg flew the South American skies, gigantic crocodiles and caimans were common throughout the tropics, and colossal prehistoric whales plied oceans worldwide alongside whale-size sharks.

This newly emerging regime of lower rainfall gave us cacti. Scientists studying the deep history of arid lands used the chloroplast genome from cacti and related species to produce a reliable phylogenetic tree, a diagram of how species are related to one another.[2] They estimate that the cactus lineage diverged from its closest relative—plants in the purslane family—about 35 million years ago in South America, long after it split off from Africa in the Cretaceous, explaining why cacti are confined to the Americas. Major diversification in the family occurred only during the last 5–10 million years ago, however. This is when the major modern cactus lineages appeared throughout the Americas.

In addition to cacti, an explosive diversification of dryland plant groups popped up throughout the world. As the cactus family was proliferating, extensive adaptive radiations were also occurring in the African, Malagasian, and American drylands. The Aizoaceae family, which includes the ice and stone plants, the aloes, the agaves, and several other succulent plant families, expanded at that time.[3] These newly appearing families share features geared toward increased water economy: a shallow root system; structures to collect and channel water to the roots; specific water storage tissues; gas exchange during the night in the highly specialized crassulacean acid metabolism (CAM) photosynthesis; thick, impermeable cuticles; and vastly modified leaves. At the same time, grass ecosystems expanded worldwide thanks to the C_4 carbon fixation that independently arose in multiple instances (more than sixty times). It offered grasses a competitive advantage over plants with the widespread C_3 carbon fixation metabolism under conditions of limited water availability, high temperatures and irradiance, and nitrogen and/or CO_2 limitation.[4] In chapter 4, Kevin Hultine explains the critical importance of these developments to saguaros.

Gradual climatic changes during the Late Miocene and Pliocene increased the drylands habitat by slowing the water cycle through increased atmospheric cooling. The increased global aridity coupled with declines of atmospheric CO_2 were the likely drivers of the diversification of succulent and grass lineages, of which the Cactaceae are probably the most representative example. We are presently in the latest phase of what scientists call the interglacial of the last Quaternary glaciation, a time of great biological and climatic changes. As global climate models universally predict sharp increases in drylands brought on by climate change, saguaros and humans alike face unprecedented climatic challenges.

For the cactus family, adapting to climate change and evolving to fit into new niches is nothing new. Over the millions of years of their occupancy of the arid and semiarid Americas, cacti have evolved into a remarkable variety of taxa. The precise number of cactus species is fluid. As researchers probe ever deeper into the molecular structure and relationships of the plants, new species emerge, and others merge and fall off the list. Arguments among researchers erupt, careers rise and fall, and the numbers of taxa change. Still, most experts agree that the number of cactus species is roughly somewhere between 1,400 and 1,850 and that there are about 125 genera. Some experts count more species, others fewer. Mexico has the most species with around 900, while Canada has until now only 4. With the advent of molecular techniques relationships among species are gradually being resolved. A recent publication called for the abolition of a large genus (*Neobuxbaumia*) and referring all its species to an older genus (*Cephalocereus*).[5] Responses to this proposal are still being formulated.

The huge gap between the numbers of cactus species in Mexico and in Canada reflects the tropical origins of the family. Mexico has the largest cacti; those in Canada are small, very small. Columnar cacti grow as far south as the province of Catamarca in Argentina, at about latitude 28° south, where we find the *cardón pasacana*, *Echinopsis* (*Trichocereus*) *atacamensis*, and as far north as Mohave County,

Arizona, at about latitude 34° north (the saguaro).[6] Outside these limits, cold temperatures appear to preclude the growth of large cacti. Cardons (which prosper in the cold of high Andean plateau country) and saguaros and a sprinkling of other large species are relatively frost tolerant, but few other columnar cacti can withstand more than fleeting encounters with freezing temperatures. At the northern and southern limits of their appearance, members of the cactus family are small, often inconspicuous, except for their spines. They seem to survive by several features, sometimes in combination. These include dehydration, nonstructural sugars, and, to the surprise of students of cacti, the depth of the snow cover.

Cacti are found in roughly six growth forms. They may be trees or at least treelike plants—arborescent and shrubby—some weighing more than 10 tons. Pereskias resemble normal trees, while columnar cacti may consist of a single enormous and lofty main stem. Other cacti may grow as vines entangled in trees, or be globular forms

FIGURE 2.1 Cactus forest in the Valley of Tehuacán, Puebla, Mexico. The dominant cactus is *Neobuxbaumia mezcalaensis*. Photo by David Yetman.

growing close to the ground. They also grow as epiphytes, like orchids, whose roots never reach the ground. Thanks to the CAM photosynthesis found in most adult cacti, they grow large in the desertlike conditions or in the canopy of tropical rain forests, where they become massive plants that, owing to their drought adaptations, are able to flourish in tree branches with little water and nutrients but plenty of sunshine. Cacti may also develop into enormous thickets, shrubs several meters tall, or diminutive ground-level plants hardly larger than a penny (approximately 20 mm). They may grow horizontally along the surface of the ground, never rising. In desert areas along the Pacific Ocean—especially in Baja California, cacti may survive primarily on fog or, in the case of numerous Chilean coastal species, on fog alone.

The Cactaceae are distinguished from other plant families by several characteristics. Most notable is the presence of spines growing from areoles, a complex structure that forms a short shoot, which we usually perceive as a felty lump from which the spines, new shoots, flowers, and fruits—namely, all reproductive structures—emerge and grow. The size, location, texture, and orientation of the areoles is often diagnostic of a species. While many plant families contain members that exhibit thorns and spines, in cacti, spines are evolutionarily derived from leaves, and they spring only from the wartlike areoles, not directly from the branches or trunks or on petioles, as is the case with most leaves. Except for a few primitive genera (that is, those appearing early in the evolution of cacti such as *Pereskia*), leaves are absent from adult cacti. Opuntioid cacti (prickly pears and chollas) exhibit rudimentary leaves early in their development but not as adults. Members of other similar plant families, such as ocotillos (*Fouquieria* spp.), live much of their lives without leaves, resembling cacti, but leaf out in response to rain or humidity, thus revealing their noncactus heritage. Some wild strains of cacti lack spines, but these are few, and they evolved in isolated environments. Hybrids or horticultural varieties of cacti have been selected to lose their spines and

are thus popular and safer in areas frequented by humans. Spines—
their size, shape, color, number, and location on the areole, texture,
and orientation, that is, their arrangement on the areoles—are often
distinctive, and thus also assist in identifying the species. Spine length
varies from more than 30 cm in some South American species to less
than 5 mm. The peculiar combination of spines and areoles usually
allows identification of the species. The spines of saguaros, for example, are usually white and located on areoles about 3 cm apart. The
spines number about thirty per areole and are often 5 cm in length,
but still, extensive individual variability exists. The South American columnar cactus *Echinopsis* (*Trichocereus*) *terscheckii*, similar
in appearance to the saguaro (often referred to as the South American saguaro), has areoles about 5 cm apart, with 5–6 spines that are
brownish or yellowish, the longest reaching 8 cm.

Recent research has revealed that spines of saguaros have rings,
similar to tree rings, that allow the researcher to trace the climatic
history that the plants have experienced during the spine development.[7] Also, as the plant grows, the succession of areoles produced
on the tip of the plant slowly migrates to the sides, producing "rings
of spines" analogous to tree rings. That linear development means
that the younger parts are at the tip, and as we proceed down the
plant, we encounter older and older tissues. This discovery, a discipline now called *acanthochronology* (Greek: *akanthos* = spine, *kronos*
= time), has become a powerful analytic tool: by studying the spines'
rings, we can determine the short-term effects of climate and biotic
agents and the long-term effects of the environment on growth and
development.

In addition to their evolutionary origin as leaves, spines provide
effective defenses for the plants against herbivores, especially mam-
malian plant eaters and large mammals that might be tempted to use
them as scratching posts or back-scratchers, as bears are inclined to
do with trees without spines. In desert climates, the spines provide

FIGURE 2.2 Variability of saguaro spines at different locations. Photos by Alberto Búrquez.

FIGURE 2.3 Spines of *Echinopsis terscheckii*, Argentina. Photo by David Yetman.

shade against sunburn and help ameliorate the effects of extreme high temperatures. In cooler locations, they help trap heat and protect against freezing. Why cactus spines exhibit such an extraordinary variation in length and composition over the family is the stuff of ongoing research. Members of the genera *Opuntia* and *Cylindropuntia* and all members of the subfamily Opuntioideae produce tiny spines called glochids, which may function to protect parts of the plants from herbivory or cold temperatures but may have additional roles, including protecting from sunburn and harvesting water from dew or fog. Glochids often appear in enormous numbers and, though somewhat inconspicuous, produce misery in creatures whose intentions were not malicious. Most, but not all, cactus fruits also bear spines, but these are usually shed or rendered inoperative when the fruits ripen and split open, thus *inviting* fruit eaters instead of repelling them.

Cacti also differ from most other plant families in that they carry on photosynthesis through their skin or the outer layers of their stems, an evolutionary development allowing them to compensate for their lack of leaves. This accounts for the green appearance of the trunks and branches. Since most, but not all, cacti live in arid or semiarid zones, having leaves would produce surfaces that increase evapotranspiration and are expensive for desert plants to produce. So, stem photosynthesis and leaflessness exemplify strategies cacti have evolved to prevent water loss. Furthermore, the outer layers or epidermis of cacti consist of thick, tough cuticle, which effectively seals most water inside the plant. Equally important, most adult cacti carry on their metabolic growth functions, that is, they produce sugars, via CAM, a biochemical process that permits uptake of carbon dioxide at night, when temperatures are cooler, instead of during the day, thereby decreasing water losses. In chapter four, Kevin Hultine discusses the importance of CAM photosynthesis in detail.

Saguaros and Rayleigh Fractionation—a Peculiar Analytic Tool

The effect of these water-conserving measures becomes apparent during the hottest and driest times of the year (May and June in the Sonoran Desert), when saguaros and other large cacti remain greenish, while other upland trees and shrubs take on the appearance of dried sticks. Since saguaros are about 80 percent water, we can accurately view them as standing drums of water that are effectively sealed from the outside. Due to their storage capacity and their ability to accumulate water from several seasons, saguaros lose component elements of water differentially by evaporation by what is known as the Rayleigh fractionation. Although a simple concept, this requires further elaboration: water is composed of hydrogen and oxygen, but these elements come in different varieties—isotopes—due to their differing numbers of neutrons. Some of these isotopes are radioactive and, with time, these unstable forms decay to stable isotopes. Oxygen has fourteen radioactive forms, none having more than a two-minute half-life, and three stable forms, ^{16}O, ^{17}O, and ^{18}O, of which ^{16}O comprises most of the oxygen on Earth (99.762 percent). Hydrogen, the other component of water, has two naturally occurring stable isotopic forms: ^1H (99.98 percent of all hydrogen), ^2H, and one radioactive ^3H (four additional isotopes do not occur in nature). The actual changes in the isotopic enrichment or depletion as compounds move between reservoirs (such as saguaros) is of great importance in climatology, hydrology, and biology. In the case of the saguaro, Rayleigh fractionation describes the process of distillation of the lighter isotopes through evaporation and the enrichment of the heavier fractions remaining within the plant, making the saguaro a reservoir of the heavier forms of water. Compared with the saguaro, the waters of the ocean and, especially, rainfall are mostly depleted of the heavy isotopic species of oxygen and hydrogen. This difference allows plant physiologists to study the sources and the use efficiency of water in producing new tissues through growth and reproduction and allows them to trace the diet sources of animals. For example, researchers have used this technique to study how much of the diet of doves came from drinking saguaro nectar and eating saguaro fruits.[8] In this sense, the saguaro, behaving as barrel, becomes a reservoir enriched with heavy water. That water, in the jargon of isotopic ecologists, is "spiked" with heavy isotopes of hydrogen or oxygen that mark the tissues of organisms consuming any part of the plant. That includes drinking their nectar or eating their fruits.

Saguaros are factories producing heavy water.

Flowers and Fruits

Most cacti produce showy to gaudy flowers that attract a remarkable variety of animal pollinators. These range from humble and industrious native bees of all sizes to moths, large and small, including the reliable hawkmoths, to beetles, to ever-present hummingbirds and bats that pollinate many columnar species. Saguaro pollination, as we discuss in chapter 3, is no simple matter. Dispersal agents for their seeds are in most cases animals that relish the pulp of cactus fruits, red, purple, yellow, and white fruits, ranging from the diminutive *chilitos* ("little chiles" that in fact resemble small domesticated chiles) of the small, globose cactus, to large pitahayas now branded as *dragon fruit* of the different species of the genus *Hylocereus*, and the succulent and flavorful red and white pitayas of many columnar cactus species. All cactus fruits are edible, and a wide variety of species yield temptingly sweet and juicy products.

Saguaros are *columnar* cacti—their trunks, and often their branches, resemble columns. The stems or trunks are strong enough to hold up the weight of the plants, which in some cases may reach many tons and, at least anecdotally, 20 m in height. Their internal structure is a

FIGURE 2.4 Fruits of columnar cacti of the Sonoran Desert. Clockwise from the top: *Pachycereus pringlei*, *Stenocereus thurberi*, *Carnegiea gigantea*, *Pachycereus pecten-aboriginum*. Center: *Stenocereus alamosensis*. Photo by Alberto Búrquez.

circle of rodlike woody ribs that provide the trunks with strength and stability. Saguaro size overlaps with the lower-size range of the largest known tree species, probably because of the increasing stiffness of mature saguaro stems.[9] When the plants die, the ribs of saguaros and many other species may persist for several years, standing or fallen to the ground, while the remainder of the plant rots away. Numerous columnar cacti possess a similar woody infrastructure, which makes the wood valuable for human use, often to the detriment of the species. Other varieties have weak ribs that quickly collapse, splinter, or rot when the plants die.

The saguaro is one of more than one hundred species of columnar cacti. They are found in every country in North and South America, except for Canada, and on nearly all the larger islands in the Caribbean. The largest ones and the greatest variety grow in Mexico. In the Valley of Tehuacán, an arid region in the southern Mexican state of Puebla and a hotbed of cactus evolution, eighteen species of columnar cacti can be found, more than in any other place. Other hotspots of columnar cacti evolution include the Balsas basin in southern Mexico and the midregions of the Río Marañón Canyon of Peru.

The saguaro survives farther north than any other columnar cactus due the combination of its ability to survive mild freezing conditions and the presence of special topographic features—the orientation of mountain ranges—that provide protection from Arctic air masses and allow the saguaro to capitalize on the passive storage of heat in their rock. Though the plants are sensitive to frosts and freezing, well-established plants appear to survive unscathed from a few hours of below-freezing temperatures. This resistance is attained mainly by their large mass that provides a lengthy time lag to reach equilibrium with ambient temperature. Species with thinner stems reach equilibrium quickly and succumb to freezing. In the northern regions of saguaros' distribution, the tips of the branches, the most frost-sensitive portions, often possess a dense mat of spines, which seems to provide the most delicate part of the cactus with considerable insulation from freezing temperatures. Southern Hemisphere species of columnars, especially *Echinopsis atacamensis* of Argentina and Bolivia, prosper in a high-altitude environment, where freezing occurs every night for months at a time. In those Andean locations, which may reach 3,500 m (11,000 feet) elevation, the great cacti produce a dense coat of protective spines, and the plants produce flowers only on the north side of the branches, where the earliest rays of the sun will melt away any frost.

Saguaros' closest relatives are included in the genus *Pachycereus*. It is a wide-ranging group, with between seven and ten members. They range as far north as central Sonora and northern Baja California, south to Guatemala. The genus includes three species that are the most massive of columnar cacti—*P. grandis* of southern Mexico, *P. pringlei*, of the Sonoran Desert, and *P. weberi* of southern Mexico, the largest of all cacti. For many years, researchers believed saguaros were more closely related to several members of the genus *Neobuxbaumia*, which has about ten species, all from central and southern Mexico.[10] Some Neobuxbaumias bear an uncanny resemblance to saguaros in shape, color, size, and flower. As Michael Sanderson relates in chapter 5, however, recent DNA studies place the saguaros' lineage closer to that of their larger neighbor in the southern Sonoran Desert, the cardón sahueso, *Pachycereus pringlei*, which is found on the central Gulf Coast of Sonora and in Baja California. This taxonomic shake-up has given rise to consternation in certain circles of students of saguaros, stoic resignation in others.

The origin of the saguaro's scientific name, *Carnegiea gigantea*, demonstrates the reach of politics into botany. The story is that the Desert Laboratory in Tucson received much of its early funding from the industrialist/philanthropist Andrew Carnegie. In 1902, while scientists and administrators at the lab were courting Carnegie's underwriting, they invited him to visit the site. They introduced him to forests of saguaros around Tucson and, thinking to flatter him, informed him that the scientific name of the plant bore his name. He wondered how such an abundant plant had eluded previous detection by such skilled botanists and was not pleased to learn that the plant's name had been changed from *Cereus giganteus* to *Carnegiea gigantea* merely to flatter and cajole him. He expressed disgust, realizing the name change was brazen pandering. He continued to fund the Desert Lab, however, and the name stuck, leaving cactus taxonomists with a heritage that would prove taxonomically complicating to later researchers classifying cacti. Botanists are subject to sternly enforced protocols to keep order in the nomenclature, one of which requires the oldest valid name for a genus be used, and thus species referred to the same genus adopt the oldest generic name. Since *Carnegiea* is older than *Pachycereus*, protocols may be invoked. *Pachycereus* may be doomed in favor of *Carnegiea*.[11]

Three

ECOLOGY OF THE SAGUARO

ALBERTO BÚRQUEZ AND DAVID YETMAN

Where They Grow

S aguaros are found only in the Sonoran Desert of the southwest
United States and northwest Mexico and its immediate environs to the south. They are a vanguard of an army of columnar
cactus species slowly moving north through the desert, thornscrub,
and dry tropical forests from their ancient origins in southern Mexico. The Sonoran is a large, hot desert covering roughly 390,000 km²
(100,000 square miles). Its north–south boundaries extend nearly
1,000 km from northwestern Arizona to Guaymas, well south of the
U.S.-Mexico border in the Mexican state of Sonora and even farther
south in the Baja California peninsula. Although most of the Baja
California peninsula is classified as Sonoran Desert, saguaros are
nowhere to be found in that peninsular desert.

Ecologist Forrest Shreve used the saguaro, along with other
representative species such as the ironwood tree and the foothills
paloverde, to delimit the boundaries of the Sonoran Desert into "phy-
togeographic" subdivisions. Today, saguaros still broadly demark the

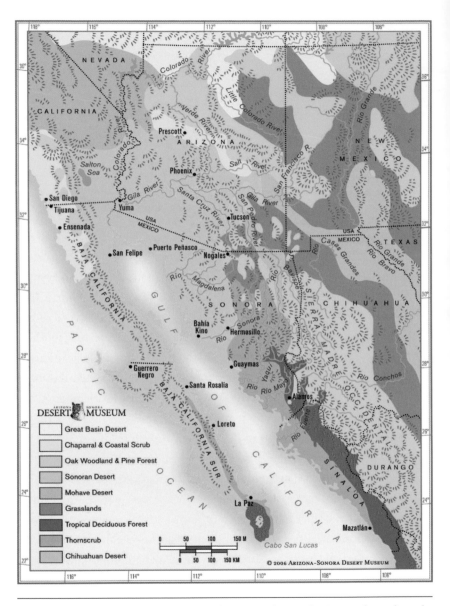

MAP 3.1 The Sonoran Desert is represented in mustard. Note that it extends nearly to the tip of the Baja California peninsula. Note also the peachy color that denotes thornscrub, or what we refer to as coastal thornscrub. Saguaros continue well to the south of the mustard. Map courtesy of the Arizona-Sonora Desert Museum.

continental boundaries of the Sonoran Desert. However, isolated populations of the great cacti grow well beyond the desert's southern limits as currently accepted by researchers. Those southernmost saguaros appear among massive boulders of basaltic lavas on Mesa Masiaca in extreme southern Sonora, and perhaps even farther south into extreme northern Sinaloa. Mesa Masiaca is a volcanic inselberg, a former lava flow mostly buried by water- and wind-borne sediments of the vast deltaic coastal plain of northwestern Mexico. The mesa juts about 125 m (400 feet) above the coastal plain and is surrounded by semitropical thornscrub vegetation. Protected within the sun-scorched lavas, saguaros can live comfortably with diminished competition from plants—especially trees—of coastal thornscrub, vegetation that is denser and more diverse than the desert scrub characteristic of the Sonoran Desert, where saguaros are more at home. In a sense, though, southern saguaro populations are indicative of biological communities at the outposts of the Sonoran Desert. In southern Sonora saguaros most likely took refuge from the cold glaciations. The populations we see now in their extreme southern distribution range are probably Pleistocene relictual populations that initiated a slow disappearance during the present interglacial period as denser tropical vegetation moved north. With global climate change, their persistence is undoubtedly imperiled.

Determining the proper location of the southern boundary of the Sonoran Desert in the state of Sonora has given rise to polite disagreement among plant ecologists. Suffice it to say that the question cannot be settled definitively, so varying accounts will place the terminus in the vicinity of Guaymas, Sonora, or as much as 150 km to the southeast.[1] Ongoing evaluations of what constitutes a desert may ultimately place the southern boundary even farther north.

Depending on one's preferences for the southern limit, additional populations of saguaros outside the boundaries of the Sonoran Desert can be found on alluvial flatlands south and east of

FIGURE 3.1 Saguaros growing on Mesa Masaica in southern Sonora. They share the habitat with organ pipe and etcho cacti, but appear to flourish on the hillside, more so than the other two species. The basaltic substrate drains quickly and heats rapidly, discouraging the more tropical thornscrub that grows at the base of the mesa. Photo by Alberto Búrquez.

Guaymas, in a narrow strip where six species of columnar cacti grow sympatrically—saguaros, *Lophocereus schottii, Pachycereus pecten-aboriginum, P. pringlei, Stenocereus alamosensis*, and *S. thurberi*. Farther south, beyond the range of *P. pringlei*, saguaros appear on the slopes of volcanic origin of the Sierra Bacatete on Yaqui indigenous lands, on the small hill called Cerro Bayájuri, west of Navojoa in southern Sonora, on volcanic rocks on the slopes of Cerro Prieto, some 20 km east of Navojoa, and on Cerro Yopori, a small hill of apparent volcanic origin some 10 km north of Masiaca. A few saguaros may persist on remote hills as far south as the state line that separates Sonora from Sinaloa, but to our knowledge these have not been documented. Small, isolated populations occur west of the Yaqui River in the foothills of the Sierra Madre Occidental, in places like San Marcial and Tecoripa.

Within the United States, saguaros are confined to Arizona except for three small populations growing in extreme southeastern California just west of the Colorado River. Farther west of those smatterings of plants, the desert is too hot and lacks the summer rains necessary for saguaros to reproduce and survive. Saguaros are also absent from Baja California, even though several species of columnar cacti thrive there, a puzzle we discuss below.

The northernmost saguaros are found on the south-facing foothills of the Hualapai Mountains near Kingman in northwestern Arizona, while their northeastern limit is foothills of the same orientation in the Gila Mountains north of the Gila River in Graham County, roughly 80 km from the state line with New Mexico. These are probably dependent on dispersal from nearby larger populations and most likely could not sustain themselves from local recruitment. At the distribution extremes of their range the plants tend to be widely scattered or isolated, unlike the denser populations—or forests—found in the central portions of the saguaro's distribution. At the colder northern limits, the plants usually grow only on south-facing slopes that feature large boulders and stone slabs, places where in cold weather the morning sun will first strike the plants, heat them above freezing, and melt any frost or snow that might have accumulated. In these colder regions, plants often take refuge on rock faces or among large boulders, which store heat and release it on frosty nights. Paradoxically, some of the largest saguaros also grow on slopes toward their northern or upper elevational limits, but local topographical features producing a mild microclimate usually explain this anomalous distribution. A gigantic plant in Saguaro National Park Rincon Unit grew above 1,300 m elevation and sported fifty-four arms. It perished in the early 1990s.

The southern Sonoran population on Mesa Masiaca concentrates on the southern and western slopes of the basaltic flow. The plants growing there are free of frosts or freezing, and the afternoon sun bakes the rocky slopes and desiccates the thin soils, conditions ideal for saguaros but difficult for their more tropical competitors. Recent research has shown that on those slopes, saguaros shirk nurse plants, which are vital to saguaro survival farther north, as we explain below.

The Saguaro and the Boundaries of the Sonoran Desert

Forrest Shreve in the 1920s suggested seven subdivisions of the Sonoran Desert based on temperature, rainfall, and vegetation. Saguaros can be found in five of these—Arizona Uplands, Lower Colorado River Valley, the Gulf Coast Region, Plains of Sonora, and in the Foothills of Sonora (no longer considered part of the Sonoran Desert), but the overwhelming majority of saguaros and the densest growth are to be found in the Arizona Uplands Division, which includes portions of northern Sonora. Examples of the greatest concentration of saguaros in this division include (but are not limited to) the Agua Fria, Hassayampa, and New River Valleys in Maricopa County; the lower Bill Williams River in Mojave County; the lower Santa Catalina and Rincon Mountains; the Tucson Mountains; the lower San Pedro River Valley; the lower Verde River Valley; and a series of low mountain ranges north and south of Interstate 8 between Casa Grande and Gila Bend, some of which are located on the Tohono O'odham Indian Reservation. Northwest of Wickenburg along Arizona State Route 93, fine stands of saguaros intermingle with Joshua trees (*Yucca brevifolia*) where the Mohave and Sonoran Deserts merge. In Mexico, extensive stands grow along a broad band from Cucurpe, located on the Río San Miguel in the east to Sonoyta in the northwest and Puerto Libertad in the southwest. Dense stands are well represented in the Caborca region, the Río Altar and its adjoining plains, between Magdalena on the Río Magdalena and the Río San Miguel, and along the Río San Ignacio, among others, all in northern Sonora. The Norwegian explorer Carl Lumholtz noted in his 1912 work *New Trails in Mexico* the existence of dense forests of saguaros in the mountains to the east and south of Sonoyta, Sonora, which is located directly across the border from Lukeville, Arizona.[2] The forests were so dense as to bring to mind "creations from the carboniferous period." Nearly all the lands of the Tohono O'odham, both those contained with the limits of the O'odham Nation and those they traditionally viewed as theirs, are thickly populated with saguaros. A survey of the photographs in the remarkable volume *O'odham Place Names*, by Harry Winters, demonstrates the vast numbers of saguaros found in O'odham homelands.[3]

Based on saguaro distribution, Shreve extended the limits of the Sonoran Desert south nearly to the state line of Sinaloa, but several authors since have noted important vegetation changes near the desert's southern periphery that lead them to reclassify the southern portion of Shreve's Plains of Sonora vegetation as Sinaloan thornscrub, which we label *coastal thornscrub*.[4] Part of the vegetation change in that newer division, that is, its divergence from Sonoran Desert vegetation, includes a greater areal percentage of ground cover and the common appearance of semitropical species reaching their northern/western limit, including (especially) the small trees *Acacia cochliacantha*, *Bursera laxiflora*, *Forchhammeria watsonii*, *Guaiacum coulteri*, *Havardia sonorae*, *Jatropha cordata*, and *Jacquinia macrocarpa*, the tree ocotillo, *Fouquieria macdougalli*, and the columnar cactus *Pachycereus pecten-aboriginum*, as well as the red-flowered small columnar cactus *Stenocereus alamosensis*.[5] Indeed, the density of the vegetation south and east of Guaymas suggests it is not desert vegetation at all. This increased density has led researchers to confine the southern limit of the Sonora Desert to the Guaymas region, or even well to the north. Saguaros are conspicuously absent from this vegetation with the exception of population pockets in especially dry microclimates and soils.

Perpetual Movement: Where They Come From and How Much Saguaros Move over Time

The distribution of the saguaro has expanded or contracted over time as the climate of the Sonoran Desert region changed, especially during periods of glaciation. Saguaros' ability to flourish during droughts and high temperatures has proven useful for dating their presence (or absence) in the northern half of the Sonoran Desert: saguaro pollen is readily fossilized in packrat middens (nests), which are permanently occupied, often for tens of thousands of years. Packrats (*Neotoma* sp.) are notorious plant gatherers, and under suitable conditions—especially those characteristic of the northern reaches of the Sonoran Desert—their urine accumulates in middens and dehydrates and solidifies over time, embedding pollen and plant parts within the dark, amber-like mass. Studies in middens demonstrate that following the lengthy cold regime of the most recent glacial period, which ended about 11,000 before the present (BP), warming began in the Sonoran Desert. This ended a long period when the vegetation was more characteristic of cool, semiarid conditions and junipers, piñon pines, and oaks abounded.

As the glaciers receded, the regional climate warmed rapidly, and saguaros (perhaps once again) marched northward from refugia to the south, reaching (and perhaps exceeding) their current range.[6] Researchers appear to have located the area where the northward march began: data by Michael Sanderson (see chapter 5) and his colleagues demonstrate that the saguaro genetic makeup can be separated into two main groups, one composed by most populations in the northern and central range in the United States and Mexico and another grouping composed of the presently highly fragmented populations in the south. Their findings also include the study of the genetic signatures of past events of the demographic history of a population at Tumamoc Hill near Tucson, a protected site toward the northern reaches of saguaro distribution. Using coalescent theory, a model of how genetic variants sampled from a population are related to a common ancestor, they merged these genetic variants into a single ancestral copy. The suggested successive colonization and retreat events of glaciers point in both cases to a refugium in the southern reaches of the present distribution range in southern Sonoran and even northern Sinaloa: saguaros found an area safe from freezing at the extreme south of their current range and, perhaps, even beyond. As mean temperatures increase in the Sonoran Desert, we can expect saguaros to be affected, though the precise dynamics of that change are subject of discussion. Upward (and northward) movement of saguaros in response to higher mean temperatures may be halted by the increasing intensity, size, and southerly sweep of frigid cyclonic weather systems of Arctic origin, which, though uncommon in the Sonoran Desert, can be expected as our planet warms. In spite of warming mean temperatures over the last century, in 2011, a prolonged spell of temperatures well below freezing from one such storm injured or killed many thousands of saguaros throughout the plants' range.

Long-term changes in the distribution of saguaros at the southern edge of the Sonoran Desert cannot be determined by the packrat midden methodology. Roughly south of Hermosillo, the capital of the state of Sonora, higher humidity precludes the desiccation needed for fossilization of pollen grains. As a result, most of the nests decompose and the pollen grains vanish.

Saguaros are of semitropical origin, meaning that they evolved under mostly frost-free climatic conditions. Their ancestors—and columnar cacti in general—evolved in the tropics, possibly from a location in the eastern Andes in South America and radiated northward, southward, and eastward, producing at least two well-differentiated clades convergent in the columnar habit, one in South America, the Trichocereeae, and one in North America, the Pachycereeae.[7] With this semitropical origin, saguaros are genetically adapted to survive only where freezing temperatures are uncommon.[8] Their large size and low surface-to-volume ratio allow them to tolerate moderate subfreezing conditions and extend their range much farther north than the other two leading northern species of columnar cactus, which have thinner stems and smaller surface-to-volume ratios, the sinita (*Lophocereus schottii*) and the organ pipe cactus (*Stenocereus thurberi*).[9] Even so, saguaro populations will experience severe die-off

FIGURE 3.2 The growing tip of a saguaro stem, showing the density of spines, areoles, and bristles. This plant, near Phoenix, grows toward the northern limit of the saguaros' range. Photo by Kevin Hultine.

during catastrophic (for them) freezes—cold spells when temperature fails to rise above freezing during the day or when, for consecutive days, nocturnal temperatures fall well below freezing, -3°C (26°F). The northern and northeastern limits of the saguaro's range are reliable indicators of locales where such cold periods and extended freezes are to be expected. Such locales are also excellent indicators of the cold-barrier limits of the Sonoran Desert.[10] A glance at the apical meristem of the saguaro, the tips of the trunk and branches where growth occurs, reveals a dense network of spines that serve, among other things, to insulate the growing tips from both freezing and sunburn, but only within certain limits. Studies have demonstrated that saguaros at the northern limits of their distribution have denser spines at their apical meristems (growing tips) than do plants farther south.[11] The saguaros' southern limits are nearly or entirely frost-free, but the increased density of vegetation as habitats become more tropical places them at a competitive disadvantage.

Climate and Its Effect on Flowering and Fruiting

In addition to their requirements for warm and desert environmental conditions, saguaros, as most other species of columnar cacti, rely on warm-weather rains, primarily monsoonal moisture, for their successful reproduction and survival. Their flowering-fruiting sequence demonstrates their deep, genetic relationship to summer rainfall.[12] Those rains usually arrive in the Sonoran Desert in the form of thunderstorms during the first two weeks of July, somewhat later in western Arizona. Students of the saguaro have long believed that mature plants did not grow as a result of winter rains, since they seem to take up little moisture at low temperatures.[13] Controlled studies, however, now demonstrate significant moisture uptake occur during the winter season rains (see Kevin Hultine's discussion in chapter 4).

During years of exceptional winter rainfall, saguaros take up so much water that their ribs sometimes burst open.[14] While the relationship of saguaro flowering to temperature has not been evaluated, studies have demonstrated for populations of organ pipe cacti significant variation in the *timing* of flowering depending on temperature: the *onset* of flowering is primarily related to the *variability* of winter minimum temperatures, the *duration* of flowering is related to the autumn-winter mean maximum temperature, and the *synchrony among individuals*—that is, the tendency for all plants to flower at roughly the same time—is well correlated with spring mean maximum temperature.[15] It is likely that saguaros respond similarly to spring and to winter variance in temperatures.

The Sonoran Desert also receives occasional late summer or fall rainstorms, often torrential, from late-season hurricanes or tropical storms originating in the Pacific. The importance of this source of moisture to saguaros' reproduction and survival is unclear, however, for the storms are sporadic and often separated by decades. But it is clear that saguaros take the opportunity to harvest and store water every time the opportunity arises. Most of the saguaros' growth occurs during the monsoon season, but despite the long history of studies on the growth of saguaros, little is known of the detailed growth rates within years and seasons.[16]

The Colorado River marks the usual western limit of monsoonal rains in the Sonoran Desert region. This climatic limit and the general lack of summer rainfall help explain saguaros' otherwise perplexing absence from the Baja California peninsula, where the closely related (and generally larger) *Pachycereus pringlei* is widely distributed. The sahueso is much more efficient in taking advantage of the scant precipitation and water brought by the fog that frequently covers large portions of the central and northern reaches of the peninsula, as we discuss below. In contrast, the saguaro seems poorly adapted to incorporate moisture from fog and dew, and since monsoonal moisture is scant in Baja California's central and northern reaches, the areas there where saguaros might be expected to grow are too hot and dry for them to survive. The requisite warm-weather moisture almost never appears there in sufficient quantity at the time when seeds are maturing. In the far southern portions of the peninsula, where monsoonal rains do occur, competitors for moisture and sunlight give saguaros little room for survival. In that region the vegetation has a thicker, more vigorous structure than the remainder of the peninsula, similar to that at the southern limits of the saguaro's distribution in Sonora, where saguaros disappear and are replaced by the more tropical etcho cactus, *Pachycereus pecten-aboriginum*, which also abounds in southern Baja California along with thick populations of the organ pipe cactus.

Saguaros are absent from the other three deserts of the United States: these dry lands are too cool or cold and lack the summer moisture that saguaros must have. Giant cacti do not appear in the Mojave Desert (cool winters with freezing days and lack of summer rains), the Great Basin Desert (cold winters and sparse summer rains), and the northern reaches of the Chihuahuan Desert (moist summers but cold winters). Beyond the southern limits of the Sonoran Desert, saguaros, beautifully adapted to dryness and water economy, cannot tolerate increased rainfall and humidity and are unable to compete with plants that thrive there, especially those with tropical affinities. Along the mainland coast of the Sea of Cortés south of Guaymas and the Pacific coast south of that, coastal thornscrub and tropical deciduous forests form a nearly continuous belt as far south and east as Costa Rica. The density of that vegetation, especially during the usually intense summer monsoons, and saguaros' inability to compete in that environment, explain their absence south of the Sonoran Desert. Many columnar cactus species thrive among the trees of those tropical habitats. Within tropical deciduous forest, however, few species attain the corpulence of the saguaro. As rainfall becomes more dependable, the need for water storage diminishes, and thinner forms with larger surface-per-unit volume become dominant inside the forest. In these southern locales, more massive cacti with relatively low surface-to-volume ratios are usually confined to steep slopes with shallow, rocky soils or microhabitats with particularly dry conditions.

In addition to rainfall and temperature requirements, saguaros in the Arizona Uplands prosper best in well-drained, rocky habitats with coarse and thin soils. They may also be found in sandy soils in warmer areas, as seen in dunelike conditions near Los Vidrios in the northern Sierra Pinacate in Mexico, just south of Organ Pipe Cactus National Monument. They occur in large numbers even in silty soils such as found in the Salt River Valley near Phoenix, and in the deep alluvial valleys of central, coastal Sonora. The greatest concentration of plants in the northern half of the saguaro's range, however, is on south- and east-facing, gentle- to steep-sloping, rocky hillsides and to the east of 112.8° W.[17] Such conditions abound in the Tucson Mountains, where saguaros appear in huge numbers on steep slopes of rough volcanic rock.

FIGURE 3.3 Saguaros on south-facing slope, Tucson Mountains. The variety of sizes and the abundance of younger plants show a favorable recruitment history. Photo by David Yetman.

Farther south, where frosts are less likely and less intense, saguaros achieve high densities on bajadas—gradually sloping montane deposits where alluvial fans produce crisscrossing drainages and water easily percolates. The saguaro's root system, as in most columnar cacti, consists of a relatively short spike root anchoring the plant and an extensive superficial root system radiating from the main trunk also functioning as supports. Saguaro, with its peculiar adaptation for intercepting water, benefits from soils where water can easily penetrate and where roots can radiate from the base, even if this requires their following cracks in rocky substrate. Saguaros can grow well on flats, as many do in central Sonora, but not in the numbers found on rocky slopes.[18] In coastal Sonora, in addition to the coarse, rocky soils, those preferred by saguaros contrast strikingly with those preferred by the massive cardones sahuesos, commonly known as sahuesos, which may do equally well on rocky slopes and on sand dunes. The largest sahueso plants grow on dense, often compact, fine-particled soils on bottomlands, even on those prone to sheet flooding. These giants send out enormous radial roots that, in addition to acting as water absorbers, function as buttresses, gigantic appendages that reach out in excess of 10 m (33 feet) from the central trunk, compensating for the lack of deeply penetrating taproots that the fine-particled soils high in clay preclude. Saguaros and sahuesos mix, but one usually greatly outnumbers the other. Sahuesos also grow best where their massive size and vast surface area enable them to benefit from high coastal humidity. During the cool nights, heavy dew condenses on the spines and cuticle and trickles down the stems to water the roots. This mechanism enables sahuesos to survive well into Baja California, in areas too dry for saguaros. On some island sites protected from grazing or predation, sahuesos achieve population densities that provoke incredulity in human observers. Saguaros are seldom found in these nearly impenetrable groves and the sahuesos in these habitats do not achieve the massive size of those on flats or sand dunes.

FIGURE 3.4 Cardón sahuesos growing on Isla Almagre Grande in the Bay of Guaymas. Island populations have all size classes well represented. In coastal localities, their range and number appear to be expanding. Photo by Alberto Búrquez.

Columnar Associates of the Saguaro

Saguaros overlap in their distribution with six other species of columnar cacti: *Lophocereus schottii* (*sinita*, old man cactus); *Pachycereus
pecten-aboriginum* (*etcho*); *P. pringlei* (*sahueso*), discussed above;
Stenocereus thurberi (organ pipe cactus, *pitayo, pitaya dulce, aaqui*
[Mayo and Yaqui]); and the lesser columnars *S. alamosensis* (*sina*,
octopus cactus) and *S. gummosus* (*pitaya agria*, sour pitaya, gallop-
ing cactus). In addition, two more species of columnar cacti occur
near the southern edge of the distribution of the saguaro, *S. montanus*
(*sahuira*, mountain organ pipe cactus) and *Pilosocereus alensis* (*pitaya
barbona*, bearded pitaya). Only in a small area in the vicinity of Guaymas, Sonora, can five species be found growing with the saguaro. Plot
ting the distribution of the eight northwesternmost species can teach
us about their requirements and spatial diversity, their similarities
and their differences, and about what makes saguaros different from
the other species.

The species with greatest distribution overlap with the saguaro is
the organ pipe cactus. This many-branched cactus (it has a national
monument designated in its honor) is the only columnar cactus apart
from the saguaro that is widely distributed within the United States,
and its range there is narrow. Plants are common near the Mexican
border in southwestern Pima County, Arizona, and reach as far north
and east as the Tucson Mountains and a small mountain called Desert
Peak in Pinal County east of Interstate 10 northwest of Tucson. Dense
forests of organ pipes formerly extended south of Guaymas nearly
into central Sinaloa in Mexico, but clearing of the thornscrub, their
favored habitat, for agricultural development has destroyed most of
these once thick populations.

Organ pipes still grow in sensational numbers in pockets along
the Sonoran coastal lowlands south of Guaymas to the Sinaloan state
line in areas where clearing has not taken place. They also abound

FIGURE 3.5 Forest of saguaros and mixed columnar cacti near Las Guásimas, Sonora. In the foreground is *Lophocereus schottii*, while *Stenocereus thurberi* can be seen in the background. The young saguaros have survived without obvious nurse plants, indicating high humidity and less chance of sunburn. This population flourishes not far south of Guaymas, near the southern limit where saguaros grow on flats and in deep soils. Photo by Alberto Búrquez.

in Baja California Sur and the southern reaches of the state of Baja California. Their range extends well into the Sierra Madre of eastern Sonora, growing at altitudes where saguaros probably cannot survive, not because of the elevation, but from competition presented by the dense vegetation growth. The flowers are white. The pulp of the fruits, the size of plums, is usually scarlet, but ranges in color from white to purple to orange. The fruits are uniformly sweet and nutritious and often marketed in cities by rural collectors.

The sahueso occupies most of the lowlands of the eastern side of the Baja California peninsula. It is present on nearly all the islands of the Gulf of California but is only found along a narrow strip of

FIGURE 3.6 Forest of organ pipes, called *pitayal* near Agiabampo, in extreme southern Sonora. This forest was cut down and the ground leveled for commercial agriculture during the 1990s. Photo by David Yetman.

coastal land in continental Sonora.[19] It is patently a peninsular species that probably crossed the gulf and colonized the coast of Sonora either a few hundred thousand years ago or recently, aided perhaps by early humans. If the latter is true, sahueso populations may still be expanding in the continent. Old individuals present in the canyons in the Sierra Libre, well in the Sonoran interior, have been shown to be contemporaneous with the retreat of Seri Indians into the range during the guerrilla warfare with Spaniards around 1760.[20] It is likely that these old sahuesos are the result of the harvest and consumption of thousands of fruits carried by humans in their frequent travels to the ancestral coastal land. Some of the seeds, known to pass unharmed through the digestive system of humans and most animals, were recruited into these majestic, large sahuesos well beyond their normal distribution range.

FIGURE 3.7 Pitayas, fruits of the organ pipe cactus, *Stenocereus thurberi*. Those of scarlet color are the most common, while the purple color, called *guinda*, is rare and highly valued. Photo by Alberto Búrquez.

The etcho cactus (*Pachycereus pecten-aboriginum*) reaches its northern limit near Guaymas, though isolated populations seem to prosper on east- and south-facing mountain slopes farther north along the inland Río San Miguel and Río Sonora. It shares habitats with saguaros in a very limited area, as if the two don't quite get along. It grows in tropical thornscrub and tropical deciduous forests in Mexico as far south as Chiapas, giving it the lengthiest distribution of any columnar cactus. The extensive distribution is marked by large gaps, and it is possible that the southern and northern populations are vicariant, that is, genetically separated. Plants in the south are exclusive dwellers of the tropical deciduous forests. They become gigantic, often exceeding 10 m in height, occasionally reaching 14 m. Etchos can be found growing not far from saguaros, but this occurs primarily

FIGURE 3.8 *Pachycereus pringlei* growing near Caborca, Sonora, one of the few locations where this species penetrates inland. Seris or O'odham probably carried fruits (or possibly plants for transplant) far from the more typical habitat on the coast. Annual rainfall here is roughly 250 mm. Photo by Alberto Búrquez.

where saguaros appear on volcanic slopes rising from the coastal plains of Sonora, where etchos are common, though not abundant, from Guaymas south to Mesa Masiaca. In tropical deciduous forest of the foothills of the Sierra Madre of southeastern Sonora saguaros are absent, but etchos often appear in great numbers. They are also common in southeastern Baja California. Flowers are white and grow down the sides of the branches. The fruits are golden, resembling large ping-pong balls with a thick covering of rather weak, golden bristly spines. The pulp of the fruits is highly edible, bright red (occasionally white or purple) and invested with comparatively large seeds, which indigenous people gather, toast, and grind into flour. The trunk is capable of producing excellent lumber that is resistant to rot.

Etchos and organ pipes grow with abundance on the plains near Mesa Masiaca, the southernmost redoubt of saguaros. Residents of adjacent Mayo villages from time to time transplant young saguaros to grow in their yards as a horticultural curiosity and decoration, much as residents of Tucson and Phoenix plant organ pipes in their yards as exotic decorations. Alas, etchos cannot tolerate frosts and do not appear to survive in the Tucson area, while they grow in the Phoenix area only in protected locations. The transplanted saguaros in Mayo country seem to fare well in yards and along urban roadways.

The sinita (*Lophocereus schottii*) inhabits a few low-lying arroyos in Organ Pipe Cactus National Monument but otherwise is confined to Sonora, northern Sinaloa, and Baja California. In the Sierra Pinacate volcanic range adjacent to the Arizona border it routinely grows together with saguaros, but unlike the saguaro, it finds Baja California acceptable. It is pollinated by a single moth species that spends its entire life living in the plant. Its flowers are a light pink to lavender color, a reflection of its disdain for attracting larger pollinators since it has its resident helper. Even so, other insect visitors regularly probe their flowers. The specialized relationship between moths and sinitas is brittle, and it might easily shift to parasitism or to a more diverse

FIGURE 3.9 Etcho cactus, *Pachycereus pecten-aboriginum*. The fruits are highly edible. The spines appear formidable but are rather weak and are readily overcome for access to the pulp. Photo by Alberto Burquez.

suite of pollinators. Sinitas also sometimes flower irregularly through warmer months. More than any other of the six species of columnar cacti of the continental Sonoran Desert region, it propagates vegetatively with relish. Cut or fallen branches left on the ground usually sprout new stems quickly. They are more frost tender in the wild but, for reasons that are unclear, appear more frost hardy than organ pipes in urban settings in southern Arizona.

Four of the six columnar cacti that share habitat with the saguaro have had no problem colonizing Baja California, yet the saguaro and the small sina cactus are absent from the peninsula. On the other hand, the creeping devil (*Stenocereus eruca*), and the dwarf organ pipe cactus (*S. littoralis*) grow only near the southern tip of the Baja California peninsula, and the candelabra cactus (*Myrtillocactus cochal*),

FIGURE 3.10 The sinita (*Lophocereus schottii*), southern Sonora. This thicket grows where rainfall averages 250 mm, far more than in the northern part of the sinita's range. Southern plants also tend to have smaller "beards." The beard structure is associated with the onset of reproduction. Photo by David Yetman.

a small columnar/candelabra cactus, grows across almost all of the peninsula but does not cross the gulf into the continent.

Their Reproduction

The saguaro's flowering period has evolved to take advantage of the summer rains, which have proved reliable, though somewhat variable, for a few million years. Flowering well before the arrival of the monsoon gives time for the fruit development and the dispersal of seeds, so that they are positioned in the right spots when the dryness ends and the first summer downpours arrive. Plants usually begin to flower in mid-April and continue through mid-June. The onset of flowering, however, may be influenced by relative fluctuations in temperatures: in the unusually warm spring of 2016 the first flowers appeared in the Tucson Mountains west of Tucson during the last week in March. During that time, buds and a few flowers appeared on the west side of the range nearly ten days earlier than those on the eastern side of the range, fewer than ten km away. By mid-April, flowering was in full force on both sides. The west side of the Tucson Mountains is lower and slightly warmer than the east side, and the earlier opening of flowers there reflects this small climatic difference. In the spring of 2017, budding began in the first week of April on the west side, and in the second week of April on the east side. In 2018, full budding did not take place until the second week of May, in spite of nearly record warmth throughout the previous months. Some fruits were fully ripened by late May, however, which is earlier than normal. A similar variability in flowering and fruiting occurs throughout the range, probably owing to the swift movement of bat pollinators, the nectarivorous lesser long-nosed bat (*Leptonycteris yerbabuenae*), during their migration northwards. The first arrivals are the pregnant females anticipating the need to nurse their babies. Later comes the onslaught of unruly males.

Saguaro flowers are white and open at night, usually after 10:00 p.m., and close forever by afternoon the following day. They are large and showy, usually 8–10 cm (3–4 inches) in diameter. The blooms are pollinated at night by a variety of flying creatures, including bats, hawkmoths, and other insects, and during the day by birds—notably doves, woodpeckers, and orioles—and insects, especially bees. In spite of the nocturnal opening of the flowers and their attractiveness to bats, diurnal bees, both native and introduced, and white-winged doves, which are migratory, appear to be the most successful pollinators in the northern reaches of the saguaro's range.[21] This puzzling relationship between flower morphology and pollinators leads one to question why a plant produces only night-opening flowers specialized for bat and moth pollination when its primary pollinators are diurnal. The question can be answered partially by the saguaro's distribution, which extends north of the range of pollinating bats. A broad swath of saguaros in the northern Sonoran Desert beyond the bats' territory may owe its existence to bees. It is likely, however, that saguaros in these marginal populations pay a price for their roaming beyond the bats in terms of genetic structure and decreased genetic diversity. Bees (and doves) have home ranges far more limited than those of bats. It would not be surprising if these populations revealed slowly developing local differentiation (speciation) that later on could be erased as the forefront of the marching saguaro vanguard and the pollinating bats move north and these formerly isolated and marginal populations become mainstream.

Bats, in their migration northward, can easily travel 60 miles while foraging and probably much more when traveling in a straight line.[22] Their ability to cover long distances quickly has had a profound effect on the genetic structure of columnar cacti: the bats' wide-ranging habits homogenize the genetic makeup of the cacti and preclude strong differentiation among populations, even those that are widely separated. Bats from one location readily transport pollen to distant populations. For the related organ pipe cactus, we have found strong components of gene flow with a predominant direction from southeast to northwest and a surprising component of gene flow in central Sonora gearing toward the Midriff Islands of the Gulf of California.[23] Our preliminary research has found a strong correspondence in single nucleotide polymorphisms among populations in opposite coasts of the gulf, suggesting long-distance genetic exchange between continental and peninsular populations. In a feat of athletic endurance, some bats apparently decide to forage across the gulf. In these trips they may carry some pollen in their fur, and/or some seeds in their bellies. As the great majority of columnar cacti share the trait of bat-pollination, it is likely that extensive gene flow is prevalent in saguaros and in all bat-pollinated and bat-dispersed species of columnar cacti. Bats help shape the evolution of many cactus species.

Although the saguaro blooms are individually short lived, each plant usually is home to a host of buds—often exceeding one hundred per branch—that open one to several at a time over several weeks, an advantage enabling the plant to respond to early or late onset of summer moisture. For much of the flowering season, mature plants nearly always have at least one flower open, an ongoing invitation to pollinators that ups the odds that the plant will reproduce successfully.

Buds of saguaro flowers often develop in cohorts, that is, buds appear in groups erupting simultaneously. Sometimes another group of buds appears a few days to weeks later, but even so, flowering is never profuse, and finding more than three open flowers in the crown of a given saguaro in a single night is unusual. Some plants may produce as many as seven cohorts per year.[24] In saguaros' northern reaches, buds of one cohort may fail entirely to bloom, while those of a later or earlier cohort may successfully bloom and set fruit. A series of unusually cool nights during April may result in failure in the cohort under development at that time. The cohort's arrested development, the time of its emergence, and the smaller number of buds may leave

FIGURE 3.11 Saguaro flowers seen from above. Some flowers have already dried up, suggesting a failure of pollinators. They open sequentially, several from each branch opening each night. Photo by Kevin Hultine.

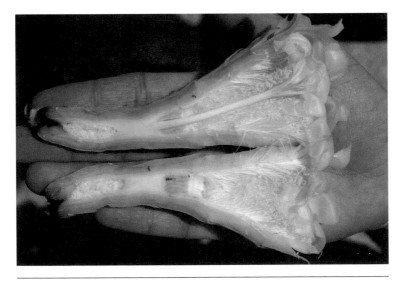

FIGURE 3.12 Cross section of saguaro flower. This demonstrates how effectively the flowers require pollinators to pass through pollen to gain access to the nectar, which lies in the bottom. Photo by Alberto Búrquez.

them increasingly vulnerable to herbivory by arthropods, especially the leaf-footed cactus bug *Narnia femorata*. A single piercing bite of *Narnia* may prove fatal to a developing bud, and bites may extend to an entire cohort. The saguaro's reproductive strategy appears to involve issuing successive cohorts of buds, thus increasing the chances of reproductive success, especially given the unpredictability of summer monsoons and the life cycle of the cactus bug. The earliest cohorts may serve as "sacrificial goats" to herbivores and cold temperatures in the same way that the sinita (*Lophocereus schottii*) is willing to sacrifice some of its fruits to the larvae of the moth that pollinates its flowers.[25]

It is likely, however, that the unpredictability of the arrival of pollinators plays a more important role in the saguaro reproductive success. We have demonstrated this for the organ pipe cacti, which often fail to produce fruits in their southern populations at the beginning of the season. The main cause for failure of organ pipes in setting fruits is the absence of their major pollinators —bats. Birds and nocturnal moths are not able to compensate for the bats' absence, and this is recorded in reproductive failure or success. At the end of the flowering season, bats arrive in significant numbers and fruit-set jumps to high levels.[26] And bats increasingly appear to hang around longer than they have in the past. We have both observed bats visiting hummingbird feeders in saguaro habitat and lingering until early fall, and some residents of the Tucson area note that in some years, bats are apparently overwintering by relying on nectar feeders, a marked recent extension of their historical visitation schedule to the northern Sonoran Desert. Since this delay in the bats' migration southward appears to be correlated with longer summers related to global warming, we can only speculate the effects on vegetation and bat reproduction strategies.

Over a sixteen-year period in William Peachey's study plot in the Rincon Mountains toward the upper limit of the saguaros' range, the average saguaro produced fifty blooms each year, suggesting that most buds fail to bloom. Single-stem saguaros have limited reproductive output, and the plant's only means of increasing fecundity is to rely on the production of side branches—arms. This growth behavior in turn increases the number of areoles where flower buds form. Branch numbers in mature plants vary considerably. Since both the main stem and number of branches varies from zero to roughly fifty, the average number of flowers per plant is wildly variable.[27] But for the saguaro in general, more branches equal more successful seed production.

Finally, individual reproductive output of a saguaro and, ultimately, its reproductive success greatly depend on the amount of photosynthates available for growth and reproduction and on the trade-off decision of the plant to allocate between these two activities: to grow bigger and taller or to produce more branches. A simple but powerful model of partitioning growth and reproduction was offered by Steenbergh and Lowe in 1983.[28] They assumed that growth from germination to late juvenile stages was not constrained and that the onset of reproduction forced the plant to devote an increasing percentage of available energy to fruit production, which diminished the individuals' rate of growth. Saguaros are forced to make tough choices.

Saguaro buds and flowers are usually located only at or near the tips of trunk or branches. Plants seldom produce branches for the first fifty years of their lives. After the first branch emerges, however, they produce additional branches more quickly, and the number of flowers increases dramatically, an indicator of the importance of branching to reproductive success. In comparison, the very tall columnar cactus *Neobuxbaumia mezcalaensis* of southern Mexico is a single stalk, but it flowers along the upper third of that stalk rather than solely at the tips, giving it more opportunities for reproduction. (Several related columnar cacti in the same region as *N. mezcalaensis* demonstrate branching patterns similar to that of the saguaro. All produce buds, flowers, and fruit at or near the apices of the branches.)

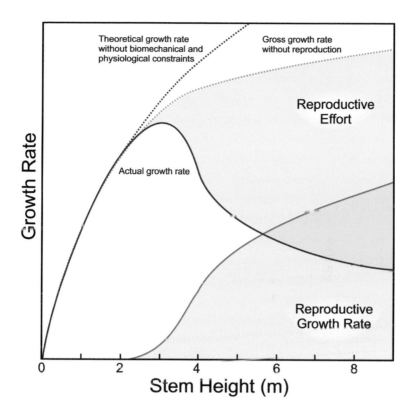

Theoretical growth rate
without biomechanical and
physiological constraints

Gross growth rate
without reproduction

Reproductive
Effort

Actual growth rate

Reproductive
Growth Rate

Growth Rate

0   2   4   6   8

Stem Height (m)

FIGURE 3.13 Trade-off model between actual growth rate and reproductive effort in saguaros. The upper blue dotted line shows a theoretical growth rate in which no energy and matter are allocated to reproduction and/or energy and matter are invested to counteract biomechanical and physiological constraints. The dotted green line shows stem growth rate, taking into account the faster increase in diameter relative to height during growth as predicted by biomechanical models. The solid blue line shows the actual growth rate reaching a maximum in the transition to adulthood—individuals 2–4 m tall—and an ever-increasing reproductive effort (shaded as light green). Reproductive effort increases at the expense of the stem growth rate. The intersection of decreasing actual stem growth rate and increasing reproductive growth rate (shaded in light blue) marks the addition of branches that soon become reproductive in older adults. (Drawing by Alberto Búrquez, modified from Steenbergh and Lowe 1983.)

Studies in the early 1960s found that from flowering to fruit ripening usually required somewhat over thirty days, a number that varies as a function of warmer or colder temperatures prior to and during the maturation period.[29] Saguaros usually set fruit in mid- to late June and early July, just prior to the onset of monsoon rains. In this way, when their seeds reach the ground—whether in fallen fruit or in the feces of consumers—they have the best chance of encountering moist and warm soil conditions, which they require for germination and growth. In 2016 fruits were normally abundant. By the fourth week in July nearly all fruits had disappeared from saguaros in the Tucson Mountains. In 2018 the first ripe fruits appeared in late May. The plants offer their sweet, red-fleshed fruits in exchange for the services of seed dispersal that are successfully accomplished by bats, woodpeckers, house finches, orioles, and many other perching birds. Doves, especially white-winged doves that consume fruits in great quantities, were long thought to be dispersal agents, but researchers have found that most, if not all, of the ingested seeds are destroyed in their gizzards, making doves formidable seed predators.

Favorable conditions for survival of large numbers of seedlings occur infrequently. Fruits are usually abundant—one study found that each saguaro produces 150 fruits each year on the average.[30] They contain dark red pulp laden with an average of more than 4,000 seeds and are contained by a rind that remains green on the exterior until maturity, when it turns reddish pink.[31] Successful recruitment, however, appears to occur only in years when timely rains are abundant and continue throughout the summer and probably extend well into autumn and winter. Such conditions are ideal for mass germination and growth of associated summer-rainfall plants that can shield the seedlings from harm, both from predation and from excessive sunlight. For optimum recruitment, successive years of above-average summer rains should occur. Fruits may be abundant and seeds prolific, but they provide food for a wide variety of creatures—mammals, birds, and even some insects. Coyotes and humans are especially fond of the fruits. During the fruiting season the desert floor is often littered with coyote feces of a reddish hue filled with undigested seeds. Compared with birds and bats, however, coyotes offer a limited range of saguaro propagation. Airborne creatures, especially bats, distribute seeds over a wide range through their feces, although many seeds find their fate sealed when deposited inside their roosting caves. The combined forces of frugivores and omnivores are quite capable of consuming a sizable proportion of the harvest of available fruits or germinated seeds, all or nearly all the fruits. Most years they do.

The necessary conditions for germination of seeds and survival of seedlings seldom occur, perhaps ten times per century, often less frequently still. The lengthy lapses between conditions conducive to recruitment lead to two observable phenomena: the appearance of many saguaros of roughly the same size, and the apparent absence of mature saguaros in some saguaro habitats. A massive die-off occurred

FIGURE 3.14 White-winged doves are an important predator of saguaro seeds. They are acrobatic feeders as well. Photo by William Herron.

at the eastern unit of Saguaro National Park when between the time of establishment of the park in 1936 and the initiation of controlled studies of recruitment in the 1960s, nearly all the mature saguaros succumbed to old age and what was once a saguaro forest became a comparatively uninteresting and structurally nondescript desert landscape. Their disappearance led many observers to lament the imminent extinction of saguaros. Subsequent studies have demonstrated that the absence of mature plants following the demise of one ancient cohort was more a function the death of a host of old plants combined with several decades of climatic conditions not conducive to recruitment and/or of management conditions that precluded the survival of younger plants, notably livestock grazing and the cutting of nurse trees for firewood. Since that time, and with the elimination of livestock grazing and firewood collecting from Saguaro National Park, a new cohort of saguaros is approaching middle age with the anticipated appearance of reproductively mature plants not far in the future. At the same time the population in the western portion of Saguaro National Park is thriving, with a host of plants now mature and a robust cohort of younger plants slowly appearing.[32] A 2004 study of thirty populations in Arizona found that each included a substantial cohort of young saguaros.[33] However, we are far from understanding the details of recruitment and survival and their causal correlates. After an exhaustive study of a large tract of saguaro habitat, another study concluded: "Just as it was impossible to know in 1960 that three decades of regeneration lay ahead for this population of saguaros, it is not possible now to know what the next century will bring."[34] Another study, examining a broad variety of saguaro habitats found that high July temperatures are a limiting factor in saguaro survival—the higher the maximum temperature, the less likely the young plants are to survive. As mean summer temperatures continue to rise in the Sonoran Desert, younger populations may be jeopardized.

Many animals compete for saguaro seeds and seedlings, as well as for the fruits. The Sonoran Desert teems with rodent and ant species, both of which consume enormous numbers of seeds and seedlings. Where overgrazing by livestock is pronounced and ground litter is largely absent, termites will invade plants, consuming soft tissues around the spines and sometimes dry fruits and fallen flowers but never healthy plants. In their classic study Steenbergh and Lowe estimated that throughout its lifetime, a typical saguaro will produce forty million seeds. Only a tiny fraction of these remain on the ground long enough to germinate, and only a tiny fraction of those that germinate survive past the first year—perhaps one surviving plant per plant. This

FIGURE 3.15 A saguaro seedling magnified twenty times. Photo by Alberto Búrquez.

is true in spite of impressive seed fecundity: under proper conditions, with ample moisture and ideal temperatures, the germination rate of seeds is in excess of 90 percent.[35] The germinated seeds provide a succulent sprout, and herbivores can make short work of the seedlings unless the numbers of sprouts exceed the appetites of herbivores or some germinate in environments protected from herbivory—or trampling. To the vagaries of weather we should add the differential chance of survival for seeds ending up in an exposed place rather than under the protection of a nurse plant.

Their Growth

From the standpoint of human longevity, saguaros in nature grow with excruciating slowness during the first few years of their lives, far slower than most columnar cacti. After five years, they may be no more than a few centimeters tall. After twenty-five years, they usually measure less than a meter in height. In moister portions of the Sonoran Desert they grow faster than in drier regions, but even so, in the wettest habitats they seldom send out branches until they reach fifty or more years of age, and the drier the environment, the longer the wait. In other words, a saguaro seed that germinates at the time of the birth of a human will usually not become reproductively active until that human reaches middle age. Unlike humans, however, the small saguaros usually grow faster just prior to the emergence of the first branch (reproductive maturity). As they reach their reproductive maturity and branch readily, the speed of growth diminishes because resources must be partitioned between growth and reproduction. (See figure 3.13, page 63.) Recently, researchers have determined that variation in temperature during the winter-spring, rather than precipitation, is more closely associated with growth than winter-spring precipitation. Increases in the seasonal variance in temperature at the beginning of the warm season result in diminished growth of saguaros. Since global climate change models uniformly project increase in both mean temperature and temperature variability, these findings do not bode well for the future of the saguaro.[36]

The onset of increased rates of branching may represent an opportunistic response to a period of unusually abundant rainfall. Branches usually emerge 2–3 m above the base, some even higher on the trunk. Saguaros in the wetter and cooler portions of the Sonoran Desert average more branches than those in the drier and hotter areas. (The 54-armed saguaro mentioned earlier grew at about 1,300 m.) In the driest and hottest portions of the saguaros' range, that is, extreme southwestern Arizona and northwestern Sonora, plants may not branch at all, and, when branching does occur, the branches tend to emerge higher on the plants. Branches of saguaros grow outward and upward but reach the height of the central trunk only if it is injured or if growing conditions subject the plants to severe water deprivation and extremely high temperatures. If branches suffer freeze damage they sometimes will droop over time, producing an appearance of downward growth.

The oldest saguaros may reach two hundred years of age, but most die well before that, the taller ones often victims of toppling by winds or lightning strikes, both of which are associated with monsoonal thunderstorms. In the colder portions of the northern Sonoran Desert, catastrophic freezes are more likely to cause saguaros' death and can affect entire cohorts of plants. On the other hand, John Alcock, who monitored the same saguaros through the year over many years' time, found that a mechanical injury as simple as constant rubbing by a palo verde or ironwood branch could render a saguaro susceptible to disease and ultimate rot.[37] Saguaros usually die slowly: they may remain upright and continue to produce flowers and fruits for up to nine years following a lethal freeze. The final stages of their death may occur with dismaying rapidity, for what recently appeared to be a healthy plant may, over a period of a few weeks, turn yellow, then brown, then brown with black streaks, branches will collapse and fall, and finally the sickly outer layers will slough off. For plants downed by lightning or wind, they may continue to produce flowers and fruits even after they topple. Their replacements are slow growing—it takes the span of an average human life to produce a mature saguaro—and the new generation may be hardly visible: when the last of the older generation dies, the younger plants may be inconspicuous, leading observers to pronounce the landscape empty of saguaros, as happened at Saguaro National Park in the 1960s. Unless housing or industrial development, roads, or heavy livestock overgrazing have destroyed the habitat, however, the young individuals are probably there, hidden in the vegetation. In a few decades, they will be highly visible.

Saguaros may weigh more than five tons. Along with other columnar cacti, they support their bodies with woody ribs, in the case of the saguaro twelve to twenty-five of them that form a circle around the developing central stalk or trunk. As Kevin Hultine explains in chapter 4, the tough exterior skin or cuticle between the ribs is pliable and capable of expanding and contracting. Part of the saguaro's mystique is its ability to swell or shrink, depending on the relative state of drought in its environs. Drought is part of the saguaro's life story, but in times of severe drought, that is, when a failure occurs in monsoonal rains, usually meager at best, the plants shrink noticeably in girth. When rain falls, the plants may absorb huge amounts of water through their widespread roots and almost overnight may be transformed from skinny, almost wizened poles to fat columns as they take in water. Their color gradually changes from a yellow green to a fresher shade of lime green. This ability to expand with water uptake is thanks to the peculiar structure of the stems (trunks). In addition to their cuticle covering, the ribs in the circular column are joined by intervening tissues that with water uptake are forced apart, superficially similar to the expansion of an accordion. The woody ribs that provide strength to the trunks are separate in younger portions of the plants but gradually fuse together as they and the plant age.

During their lifetimes, saguaros host a variety of birds in their trunks, probably reluctantly. Most commonly these are Gila wood-peckers and gilded flickers, both of which use their powerful beaks to hollow out sites for their nests on the thick trunks and branches. These birds may play a role (certainly a minor one) in pollinating the flowers and dispersing the seeds and may also protect the plants by harvesting potentially harmful larva from diseased tissue around the nest sites. The saguaros respond to these woodpecker attacks by form-ing a dense layer of callous or scar tissue around the cavity, popularly referred to as a boot.[38] Once the woodpeckers have abandoned the site small owls or kestrels often occupy it.

Ecologist John Alcock observed that gilded flickers drill their nest holes higher on the branches and trunks than those hollowed out by Gila woodpeckers. He attributed this to the flicker's narrower beak, which forces the bird to seek cavity sites with softer tissues—only available on the younger, higher portions of the trunk or branches. The flickers' preference produces more evident damage in saguaros—their activity may weaken the top, less seasoned section of the trunk or branch, causing it to fold over. Whether or not the woodpeckers' attacks also weaken the plants, the trunks accumulate woodpecker-nesting cavities as they near senescence, or perhaps old plants are more inviting to woodpeckers. A large saguaro with an abundance of nesting holes, especially on its upper portion, is usually in the early stages of death. When the old giants topple and rot away, the boots formed by the hardened tissues often survive as ghostly relics among the ruins of the once imposing plant eerily outlined by the unique minerals solely found in saguaros and a few scattered spines.[39]

The stems and branches of saguaros and most cacti seem to be largely immune to herbivory. This is due to the proliferation of spines that deter herbivores and to the high concentration of oxalates and

FIGURE 3.16 Dead saguaro. The skeleton several years after the plant's passing.
Photo by Alberto Búrquez.

other secondary compounds in the tissues that are either toxic or
unpleasant to herbivores. Packrats may develop a tolerance for oxalates, and in times of extreme drought, jackrabbits seem to gnaw at
plant bases, much to the detriment of the saguaro. This destructive

action is especially notable among saguaros in extreme southwestern Arizona, the driest part of their range.[40]

Their Germination and Establishment

Germination of seeds marks the start of what may be a life of a couple of centuries. The odds of this happening, however, are infinitesimally small, for even those few individuals that survive into adulthood seldom survive the onslaught of natural dangers. Since the work of Forrest Shreve in 1910, researchers have noted the effect of trees—especially foothills palo verde—in sheltering saguaro plants from the intense desert irradiance. With time, this observation led to the hypothesis that the saguaro owed its recruitment success to the presence of desert trees that buffered the physical and biotic environment, thereby increasing their chances of survival, hence the term *nurse plants*. This spatial association was viewed by many ecologists as an idiosyncratic feature of the paloverde-saguaro relationship and not an example of a general principle of positive associations in harsh or stressful environments.[41] Today, most ecologists agree that positive associations play as big a role in structuring ecosystems as other symbioses, such as competition or predation. Many plants, as well as physical features of the environment such as rocks and crevices, provide sheltering places that vastly enhance the chances of the recruitment and survival of other species. Most saguaro seeds are consumed by granivores—seed predators like doves—and are deposited in unsuitable places like rock surfaces and in deep caves where bats roost, or they rot away before they have a chance to germinate. Those seeds that are fortunate enough to find moist and warm soils and succeed in germinating produce two tiny cotyledons, succulent leaves that are quickly gobbled up by all manner of eager creatures, especially rabbits, rodents, birds, and ants. These minute, tender primary leaves are also easily trampled, especially by livestock and deer, or desiccated, if they are exposed to direct sunlight. When temperatures fall below freezing, they lack sufficient mass to prevent their death. Their chances of survival in colder parts of the Sonoran Desert are vastly enhanced if they find refuge under nurse plants, probably essential in the northern half of the saguaro's range. Nurse plants are so named because their shade, branches, and litter, plus the other plants that grow in their shade, shelter the seedlings from sunburn and freezing, protect them against trampling, conceal and shelter them from herbivores, and in the case of some leguminous trees, provide a nitrogen-enriched soil environment. Nurse plants range from the small shrub triangle leaf bursage (*Ambrosia deltoidea*) to desert trees.[42]

In Arizona, the most vigorous young plants will be found growing beneath foothills palo verde (*Parkinsonia microphylla*) and ironwood (*Olneya tesota*), especially the former.[43] In the case of the

FIGURE 3.17 Young saguaros growing underneath a nurse tree, in this case a mesquite. Cucurpe, Sonora. Photo by Alberto Búrquez.

palo verde, as the saguaro ages, its shallow, radiating roots begin to intercept moisture needed by the tree, and it may ultimately kill the very plant that gave it nurture and provided for its survival. This cycle occurs so slowly, however, requiring decades for completion, that regeneration of replacement palo verde occurs, maintaining a relatively constant population of the trees. Once saguaros are large enough to endanger the life of the nurse plant they have achieved body mass sufficient to withstand all but the most intense freezes found in the Sonoran Desert.

In the driest portions of the Sonoran Desert, the common creosote bush (*Larrea divaricata*) and triangle leaf bursage (*Ambrosia deltoidea*) sometimes serve as nurse plants. Our ongoing research shows a high dependence of saguaros on the protection of nurses in the northern and the driest reaches of distribution and a declining association with plant canopies as the desert is replaced by the thicker canopy of the thornscrub in southern and southeastern Sonoran saguaro outposts. It appears that in the northern portion of saguaro habitat, the cover of trees and bushes provide protection against freezing, solar burning, desiccation, and predation, but in the south, plant canopies become so dense and humidity so high during the wet season that young saguaro saplings suffocate. Saguaros continue to be found to the south, but only on steep, rocky hillsides, usually among basaltic boulders, which appear to replace plants as nurses for saguaros.

Senescence and Survival of the Species

It is useful to imagine a saguaro as an ever-growing, giant, prickly leaf that endures for more than one hundred years. During growth, saguaros continuously add photosynthetic tissue, which remains fixed at the growing point where it was produced. For example, a 1 m tall individual at the Saguaro National Park in Arizona will be about twenty-five years old. It will produce fresh, young, green tissue at its growing tip. A 10 m tall individual will be about one hundred years old and will continue to produce new green tissue at its tip. In this old individual, however, the once freshly green tissue at one meter will be a venerable seventy-five years old—the same tissue, at the same place from which it issued those many decades ago, but now beset with the travails of old age. As saguaros age, bark injuries, also termed *epidermal browning*, start appearing on the stem surfaces, resulting in blackish, rough skin. This change in color and texture differs markedly from the typical green skin, which includes protective layers of cuticle, epidermis, and hypodermis, along with the green photosynthetic tissues deeper into the chlorenchyma. Younger saguaros and areas with new tissue have a smooth, green surface on which CAM photosynthesis occurs, but as in humans, age leads to the accumulation of imperfections and blemishes from a variety of hazards—animal predation, physical injuries, and attacks of microorganisms. Bark also recapitulates a history of the deleterious effects of UVB irradiation, as does human skin. The bark, gray/black in color, begins forming on south-facing surfaces. It gradually spreads around the saguaros' lower body, often completely covering the stem perimeter, a process that in vulnerable individuals begins around the young age of forty-five. Since it cannot carry on photosynthesis, the encroaching dark bark catches up with the ability of the saguaro to grow, often resulting in increased mortality rates among adult saguaros. Saguaros outlive us humans, but they still suffer the ravages of time and relentless savaging by the sun. Inevitably, the plants succumb to old age. The leading causes of premature saguaro death—lightning strikes and high winds—may present a quicker and more compassionate exit from the Sonoran Desert than slow demise from epidermal browning.

The survival of saguaros as a species seems secure far into the future. Threats to populations linger over large swathes of saguaro habitat, however. They are fourfold:

First, they face an onslaught of habitat destruction, especially in Mexico. Most of the densest populations within Arizona lie within protected areas—national and local parks, wildlife refuges and national forests. Urban expansion has consumed large tracts of land, especially in the Phoenix metropolitan area. Arizona statutes prohibit arbitrary killing or destruction of saguaros, but many thousands of acres of habitat have been given over to urban development in the last fifty years, and enforcement of prohibition on removal is sporadic. Transplanting of young saguaros is usually successful. Once the plants have branched, however, transplanting is only marginally successful.

In Mexico, despite being listed in the Mexican endangered species act (NOM 059) the plants seem even less protected. The original construction of Mexico Route 15, connecting Nogales, Sonora, with Mexico City, was routed through one of the densest forests of saguaros, resulting in the bulldozing of thousands of plants. Expansion of the freeway in 2016 multiplied the insult to one of Sonora's finest populations. Overstocking of livestock on ranges continues to pose a major problem for recruitment, as does the rampant collection of dead skeletons for the saguaro rib trade.

Second, invasive grasses have proved deadly to saguaros and their threat is expanding. The most insidious are red brome (*Bromus rubens*) and buffelgrass (*Cenchrus ciliaris*). Red brome is an annual grass of Mediterranean origin that evolved in a climate of cool, moist winters and hot, dry summers. It germinates in the fall and with even marginal winter rains can quickly cover large acreages of desert landscape, crowding out perennial grasses and other winter annuals. With rain, the grass sets an attractive-looking carpet on the desert floor. Once seed is set, however, the plants dry quickly and become a fire hazard while the sharp awns that protect the seeds attach readily to fur and clothing. The seeds are not affected by brush fires and will survive to germinate the following year, but the desert plants where the red brome flourishes are not accustomed to fires. They and their seeds are consumed by fire. Red brome is an especially urgent problem in the moister desert habitats in Arizona, where range fires fueled primarily by red brome in recent years have destroyed many thousands of saguaros and decimated populations of other desert plants.

Buffelgrass is even more noxious to native Sonoran Desert vegetation. A perennial African grass deliberately introduced by the U.S Department of Agriculture in the 1970s to improve forage for cattle production, buffelgrass is extremely well adapted to Sonoran Desert conditions. The seed germinates readily, especially in disturbed soils, and forms dense clumps of grass. When the grass dries out, it becomes readily flammable, seemingly inviting fire. While the fire kills native Sonoran Desert vegetation, buffelgrass evolved on savannahs, where fire is part of annual cycles and the flames stimulate the grass to rapid growth. Over time fire consumes much of the native vegetation, while buffelgrass flourishes, producing in many areas stands of nearly pure grass, pastures that support greatly reduced plant and animal diversity. It has spread rapidly throughout the southwestern United States and northwestern Mexico (and many other areas in the world), constituting a short-term boon to some ranchers but immense destruction to biological diversity.

Third, climate change poses several threats, mostly conjectural now. Studies by scientists at the Tree Ring Laboratory of the University of Arizona have revealed that climate change has been pushing the tropics north in the Northern Hemisphere, up to latitude 4° (about 450 km) in the last eight hundred years. Climate warming should enable saguaros to grow higher on hillsides and farther to the north of the present limits. Arctic storms, however, may grow larger as heat produces larger changes in the oceanic circulation, and the reach of freezing temperatures and its variance may remain constant or even expand southward as the planet warms. For example, hard and prolonged freezes in 2011 and again in 2018 took a heavy toll on young and old plants in southern Arizona and northern Sonora. If this pattern becomes the rule, saguaros may have to battle to maintain the northern limits of their range. Increased prolonged high temperatures, already occurring in the saguaros' range, may also increase mortality from sunburn and desiccation among mature plants.[44]

Finally, the incorporation of the dried ribs from dead saguaros into furniture and luxury design themes in affluent housing in the southwestern United States has a potentially drastic effect on saguaro ecology and even on living populations. Though dead as well as live plants are protected with parks and monuments, the ribs' value encourages poaching and killing mature plants. In Mexico, plundering of plants, live and dead, for trade in saguaro ribs proceeds virtually without government intervention.

FIGURE 3.18 Crafts fashioned from saguaro ribs marketed toward tourist trade in Puerto Peñasco, Sonora. Photo by Alberto Búrquez.

Four

THE ANATOMY AND PHYSIOLOGY OF THE SAGUARO

KEVIN HULTINE

Giant cacti are among the most charismatic plants on the planet in terms of their physiology and morphology and are widely regarded for their cultural, economic, and ecological value. They occur naturally in some of the driest and least vegetatively productive subtropical regions of the Americas. To persist in harsh arid environments, cacti have evolved many specialized traits, including a thick, waxy cuticle that restricts excessive water loss; the modification of leaves into spines combined into short shoot systems termed areoles; and the exhibition of crassulacean acid metabolism (CAM), a carbon-concentrating photosynthetic pathway that allows plants to acquire carbon dioxide (CO_2) at night when water loss from photosynthetic tissues is minimized.[1] Among the most important features of cacti are their photosynthetic stems that in many species store massive amounts of water and other resources to support growth, reproduction, and survival during hot and dry conditions.

Giant saguaro (*Carnegiea gigantea*) is arguably the most iconic and perhaps among the most ecologically important giant cacti species. Saguaro occurs throughout the Sonoran Desert region, and its range spans a fivefold gradient in mean annual precipitation from about 100 mm (4 inches) in the northwestern Sonoran Desert to about 450 mm (18 inches) in the extreme southern edge of its range (see the previous chapter for a detailed description of the geographic distribution of saguaro). Throughout its range, giant saguaro is an important foundation species that supports numerous frugivores, nectarivores, and other fauna.[2] However, the capacity for giant saguaro to serve as a foundation species is not only dependent on rare surges of seedling establishment during moist years but also on its survival and growth through long stressful periods of limited rainfall as well. In this chapter, we focus on the physiological and anatomical characteristics that not only give saguaros the stature that makes them so recognizable but give them as well a considerable edge over almost all plant species for coping with drought and extreme aridity. We start by reviewing the important aspects of the saguaro stem, which is capable of storing copious amounts of water over long durations. Then we describe the specialized photosynthetic strategies that saguaro deploys to minimize water loss from its tissues.

Anatomical Features of the Saguaro

The cactus family (Cactaceae) is comprised of 1,450–1,850 species distributed throughout the arid and semiarid regions of the Americas. Nearly all cactus species are characterized as stem succulents (i.e., plants with green, fleshy stems). Giant saguaro is among the largest of all succulent-stemmed plants, with the tallest reaching heights of greater than 18 m (60 feet). Their highly porous tissues can have a relative water content that exceeds 90 percent by mass, thereby allowing saguaros to store massive amounts of water in their stems.[3] For example, if we assume that an entire saguaro stem is constructed of storage tissue, a 3 m tall plant (10 feet) with no arms and an average diameter of 30 cm (1 foot) would contain about 170 liters (45 gallons). This means that a 10 m tall saguaro with multiple arms like the one pictured in figure 4.1 could hold more than 3,500 liters (920 gallons) of water! In reality, a saguaro stem is not constructed entirely of storage tissue. Therefore, it is useful to identify the major components of a saguaro stem and the important functions that each component provides.

The saguaro stem has five major functions: protection, photosynthesis, storage, water transport, and structure. These functional characteristics are separated over the cross section of the stem such that the different anatomical features serve a specific function to support the growth and survival of saguaros under what are often harsh environmental conditions.

The outside "skin" of a cactus is called the epidermis. The epidermis of the saguaro, containing multiple cell layers, is thick compared to many other desert cacti.[4] The outermost surface of the epidermis is the waxy cuticle formed almost entirely of fatty acids. The cuticle prevents water loss from the cactus skin and is indigestible by small organisms that might try to enter the saguaro and grow in its succulent tissues. Small pores, called stomates (or stomata), that are too small to see without a microscope, are dispersed throughout the skin of the cactus.

Importantly, the cuticle generally does not cover the stomates, thus allowing the exchange of gases with the surrounding air. The aperture of each stomate is controlled by two kidney-shaped guard cells. When these guard cells fill with water, they become turgid and expand like balloons, thereby increasing the aperture of the stomate pore. As the stomatal pore aperture increases, the rate of CO_2 diffusion into the stem also increases. However, the increased CO_2 diffusion comes with a tradeoff in that the diffusion of water vapor from moist, internal tissues of the stem to the dry atmosphere also increases. As conditions get progressively drier and hotter, the water inside the guard

FIGURE 4.1 Photo of a giant saguaro occurring near Cave Creek, Arizona, standing over 10 m tall with eight main arms visible. Photo by Kevin Hultine.

cells is released, causing them to lose their kidney shape and flatten over the top of the stomatal pore. The reduction in pore size prevents excess water loss but also suppresses CO_2 intake diffusion, which is necessary to conduct photosynthesis. Desert plants such as saguaro, therefore, must find creative mechanisms to maximize CO_2 diffusion while simultaneously preventing excessive water loss. These mechanisms are discussed later in this chapter.

Another important function of the cuticle is preventing excess sunlight and heat load on the saguaro stem. Holding a peel of the cuticle to a lamp reveals that it is translucent and therefore permits light to penetrate the internal tissues of the cactus.[5] However, the saguaro cuticle is virtually opaque to harmful ultraviolet radiation, while permitting 70 percent of the sun's wavelengths that are necessary to initiate the biochemical reactions of photosynthesis. The exclusion of UV radiation by the cuticle not only protects the photosynthetic tissues from damage, it also keeps the massive saguaro stem from overheating. Nevertheless, saguaros still must cope with extreme temperatures, with tissues directly underneath the skin reaching 55° C (131° F) in mature, healthy stems.[6]

Immediately underneath the epidermis is the hypodermis: a layer of cells called the collenchyma that is used by saguaro and other cactus plants for mechanical support. These living cells contain a high concentration of a water-holding substance called pectin. An important feature of pectin is that it attracts water, which fills pectin cell walls and makes them rigid but flexible at the same time. The flexibility of the collenchyma allows the stem to expand as it takes up water from the soil and contract slowly over time as water evaporates through the stomata. The hypodermis layer is particularly thick (more than 1 mm thick) in saguaro (figure 4.2) and other giant cacti species in order to provide structural support to the large stems but remains flexible enough to allow the stem to undergo dramatic shrink-swell cycles over the course of the year.[7] The thick hypodermis also provides an extra layer of defense against small animals and insects that might otherwise feed on the succulent saguaro stems. Despite its thickness, however, the hypodermis is not an uninterrupted structure around the stem. Passageways called substomatal cavities develop between the epidermis and the inner tissues of the stem through the hypodermis. These passageways are the conduits for CO_2 uptake and water loss from the saguaro stem. In moderately young saguaro plants (less than fifty years old), the epidermis and the hypodermis cover the stem from the apical meristem located at the top of the stem to the ground surface. As saguaros age, the skin near the base of the stem is replaced by a thick bark layer called the periderm. As the bark forms, the thick cell walls of the hypodermis become progressively thinner, eventually becoming indistinguishable from the tissues of the inner stem.

Beneath the hypodermis is the cortex and the pith, collectively referred to as the fundamental tissue of the cactus plant. This fundamental tissue carries out two important functions related to the extreme adaptation to aridity in saguaro and other giant cacti: photosynthesis and water storage. The outer cortex—closest to the hypodermis—is called the chlorenchyma and is comprised of green photosynthetic cells (figure 4.2). It is here that CO_2, water, and sunlight combine to produce sugars from photosynthesis. The specific photosynthetic strategies of saguaro are discussed in detail later in this chapter, but it is still worth describing the important anatomical features of the chlorenchyma here. Carbon dioxide from the atmosphere diffuses through the stomate and the substomatal cavity to the cell walls of chlorenchyma cells. The CO_2 gas is then dissolved in water, where it passes through the walls of specialized cells that contain a large central vacuole, which can occupy 95 percent of the cell volume. Sunlight passes through the translucent skin of the cactus and is absorbed by the photosynthetic pigments in the chloroplasts in the chlorenchyma cells. The absorbed light energy is converted into chemical energy that ultimately leads to the production of glucose and other sugars in the chlorenchyma.

Beneath the chlorenchyma cells are the inner cortex and the pith, which have a primary function of water storage. No other anatomical features of the giant saguaro are functionally more important to

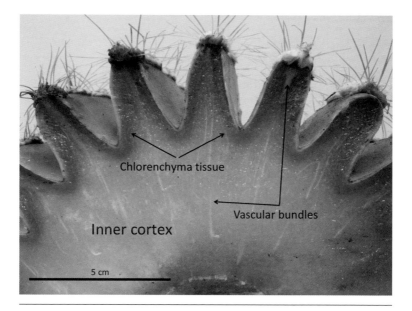

FIGURE 4.2 Semi cross section of a saguaro stem near its apex. The cross section shows the spines and areoles positioned on top of the thick collenchyma (skin) of the saguaro. The dark green photosynthetic chlorenchyma layer is clearly visible immediately below the skin, with the thick inner cortex below the chlorenchyma. Notice the vascular bundles extending through the cortex to the base of the areoles.

its survival during drought. As discussed earlier, saguaro stems can store a massive amount of water, and the primary storage reservoirs for all of this water are the inner cortex and the pith. The cells in the inner cortex are constructed so that they can retain a large volume of water in their central vacuoles. In fact, the cells of the inner cortex can hold four times more water than the cells in the chlorenchyma layer.[8] The highly flexible walls permit the collapsible cells to release water to the photosynthetic tissue in the chlorenchyma, where water is lost to the atmosphere through the substomatal cavities and the stomatal pores. High water storage also occurs in the pith, located in the center of the stem. In most cacti species, the pith occupies only a small fraction of the stem volume, but the pith in saguaros can occupy 20 percent or more of the stem. Similar to the inner cortex, the cells located in the pith have thin walls that are highly flexible, with large vacuoles for water storage.

A difficult anatomical challenge that saguaro and all other giant cacti face is transporting water from the roots upward (axially) to the top of stems, then outward (radially) to the photosynthetic tissues near the surface. One would assume that saguaros have evolved with the necessary plumbing to overcome these challenges or they would not have succeeded in obtaining their massive size and stature. Indeed, the vascular system that is constructed to transport water, nutrients, photosynthetic products, and other organic molecules involves a complex network formed from both live and dead cells. This complex network starts with the root system. Large cactus plants such as saguaro can construct new, short-lived "ephemeral roots" within eight hours after a rain, allowing rapid absorption of water from the surrounding soil.[9] After traveling radially across nonlignified (i.e., nonwoody) root tissues, the water molecules are literally pulled into tubes, called xylem vessels, either by pressure gradients that form when liquid water evaporates out of the stomatal pores (i.e., the same forces that cause soda to be sucked through a straw) or by osmotic gradients caused by high solute concentrations in the cortex and pith. Once water molecules enter the base of the stem, they travel axially through xylem vessels organized in vascular bundles, often called *axial bundles*, that are located between the inner cortex and the pith. About two hundred axial bundles exist in saguaro stems, considerably more than the number of bundles found in smaller-stemmed cacti species.[10] The large number of axial bundles in saguaro stems is in no way surprising, given the large volume of water storage tissues in the pith, inner cortex, and chlorenchyma that must be supported by a robust vascular system.

From the axial bundles, water can be diverted into one of three types of conduits: smaller bundles that diverge from the axial bundles toward the outer ribs, cortical bundles that are distributed throughout the cortex (figure 4.2), and medullary bundles that are distributed throughout the pith. One of the most important and recognizable features of any cactus plant are the specialized axillary buds called areoles that form on the outer ribs (figure 4.2). Spines (i.e., modified leaves) in saguaros and other cacti grow in clusters from areoles and tend to persist on the areole for several decades. Small vascular bundles extend from the axial bundles to the base of the areole but terminate at the base of the spine (figure 4.2). These vascular bundles do not function to directly support spine growth, but instead support reproductive structures that develop from the areoles. Water is carried to developing flowers and fruits in xylem vessels, and sugars for tissue construction are carried to the reproductive structures in even smaller tubes called phloem. This complex but efficient vascular system is necessary because mature saguaros produce on average three hundred flowers per year, and each flower can produce 1 ml of nectar in a single evening.[11] Thus, reproduction in mature saguaros comes at a considerable cost in terms of water and energy.

Cortical bundles are a complex network of xylem and phloem tissues that extend from the axial bundles to the inner cortex and chlorenchyma. They spread in all directions like leaf veins and change directions throughout but do not extend to the hypodermis. Cortical bundles have three important functions: (1) transporting water throughout the cortex, (2) transporting sugars to and from storage cells in the inner cortex, and (3) transporting products produced from photosynthesis from the chlorenchyma tissues to the axial bundles.[12] Medullary bundles are similar in size to cortical bundles and are also comprised of xylem vessels and phloem tubes. However, these bundles are found in the pith instead of the cortex and tend to occur only in cactus species like saguaro that have large-diameter piths. The primary function of medullary bundles is for the transport of water and sugars from the axial bundles to and from the pith.[13]

The complex vascular system, coupled with its large stem storage volume, allows saguaros to take up and store water rapidly, then slowly lose it during photosynthetic gas exchange throughout the year. An example of the dynamic nature of saguaro water use can be illustrated by its dramatic seasonal variation in stem diameter over the course of the year (figure 4.3). The stem diameter of this 4 m tall saguaro growing in Phoenix varied from a maximum of about 39 cm to a minimum of a little more than 33 cm. The minimum stem diameter occurred in both years in summer (Day 243: August 31, 2015, and Day 561: July 15, 2016) prior to significant monsoon rainfall. The period just before the summer monsoon is usually when saguaros are under the most stress because they have little or no soil moisture from which to extract water. Therefore, they must draw on the internal water reserves until monsoon rains recharge the rooting zone with moisture. Once the soils become wet again, saguaros take up and transport water to the storage reservoirs in the cortex and pith. As the large vacuoles in the specialized cells in the inner cortex and pith fill with water, the stem rapidly increases in diameter (figure 4.3). Between rains, the stem contracts as water is lost during photosynthesis.

Although mature saguaros only grow in summer and early fall, figure 4.3 reveals that they are still actively taking up water during other times of the year. For example, the stem expands between Day 365 (December 31, 2015) and Day 401 (February 4, 2016) after a moderate rain. The capacity to acquire and store water during the winter may have implications for summer growth, especially during exceptionally dry summers or in locations where summer moisture is far less reliable, such as the northwestern edge of the Sonoran Desert.

An important anatomical feature of saguaros and other cacti that allows the stems to expand and contract is the outer ribs. If cactus stems contained no ribs and were instead perfect cylinders, their stems would literally tear apart if they were to expand after a large rain. Instead, when a stem swells as it absorbs water, the ribs expand,

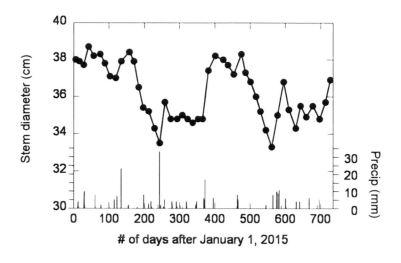

FIGURE 4.3 Stem diameter measured biweekly for two years on a 4 m tall giant saguaro (photo) growing in Phoenix, Arizona. The bars represent the amount of daily precipitation that fell between January 1, 2015, and December 31, 2016.

and their bases become broader. If the ribs could expand to the point that they become flat, the stem would in fact turn into a cylinder. In reality, the ribs are not flexible enough to expand to such an extent, and thus the stem must find other ways to maximize flexibility and support. One way to support more volume change is to construct more ribs, and indeed there is some evidence that saguaro populations in wetter habitats have on average more ribs around the stem than those in drier habitats. A greater number of ribs also means that potentially more areoles can be produced during the growing season since areoles grow from the apex of each rib. The advantage of more areoles is that each areole produced during the summer has the potential to produce a flower during the following spring.

When fully hydrated, a saguaro stem the size of the one shown in figure 4.3 can weigh close to 250 kg (550 pounds). Therefore, saguaro stems need sturdy structural support to remain upright, especially when they are filled with water after large rains. Saguaro and other columnar cactus species have internal ribs organized in a cylinder between the inner cortex and the pith. This woody matrix of vertical ribs is largely constructed of fibers with lignified cell walls. The fiber cells are elongated parallel to the axis of the stem, giving the ribs strength, flexibility, and resistance to breaking. Saguaros are unlike many smaller columnar cactus species in that as they grow taller, the woody tissues and the associated vascular system spread apart, allowing the pith to expand with age.[14] This clever growth strategy is partially why saguaro stems can attain their massive girth while simultaneously maintaining the necessary structural support to persist for many decades. When a saguaro dies, the skin peels off the stem and the succulent tissues underneath dry out and decompose. This exposes the recognizable woody skeleton that has the appearance of vertical rods organized in a cylinder. The lignified rods decompose extremely slowly in the dry desert environment, and thus the saguaro skeleton can remain standing for many years after the rest of the plant is gone.

Saguaro Photosynthesis

Before describing the nuts and bolts of the biochemical process of photosynthesis in saguaro stems, it is worth reviewing the important implications of stem storage volume versus surface area. In the most straightforward terms, volume-to-surface-area ratio is analogous to economic supply and demand. The supply of water and other resources to the photosynthetic tissues is determined by the storage capacity in the inner cortex and pith, while the demand is determined by the amount of stem surface area and photosynthetic activity. When supply is higher than demand, plants can conduct photosynthetic gas exchange at their highest capacity since water conservation strategies are unnecessary. However, when demand outpaces supply, water loss must be regulated by reducing the amount of water vapor that can escape from the stomatal pores to the drier atmosphere. In turn, this reduces CO_2 diffusion into the stem and limits the rate at which photosynthetic enzymes can convert CO_2 into sugars. Across the entire cactus family there is a two-hundred-fold range in stem volume to surface area ratios, and giant saguaro stems have among the largest volume-to-surface-area ratios among all columnar cactus species (figure 4.4).[15] Giant saguaro and other cactus species that have large stem volumes relative to surface area tend to occur in locations that are drier than those inhabited by cactus species with smaller volumes.[16] Intuitively this makes sense given the importance of maintaining a large storage volume to cope with extended periods of drought. The cost for all of this storage is a lower fraction of chlorenchyma cells where photosynthetic sugars are produced for tissue growth, reproduction, and other plant functions. This also means that saguaros are slower growing than species with more photosynthetic surface area and are therefore confined to dry deserts where they need not compete for sunlight with faster growing plants.[17]

FIGURE 4.4 Stem volume-to-surface-area ratio of fourteen cultivated giant cacti species (n = 1 individual stem per species) co-occurring at the Desert Botanical Garden in Phoenix, Arizona. The photo inset shows an example of the Desert Botanical Garden's living collections, highlighting a giant saguaro and a senita cactus with red arrows identifying their measured volume-to-surface-area ratio.

Despite their large storage volume, saguaros still must find water conservation mechanisms to conduct photosynthesis in the desert environment. A common feature of succulent plant species such as saguaro is the evolution of a complex photosynthetic pathway called crassulacean acid metabolism or CAM. The advantage of CAM is that it increases the rate of photosynthetic carbon fixation relative to the rate of water loss (in other words, CAM limits the demand for water) but is also metabolically costly compared to the more common C_3 pathway used by most plants. With extremely rare exceptions, all cactus species use the CAM photosynthetic pathway. However, there are important variances in how the CAM pathway functions in different species. In order to describe these variances, we must first compare and contrast CAM with the more common C_3 photosynthetic pathway.

Plants use chemical energy acquired from sunlight to produce sugars from CO_2. For the large majority of plant species on Earth, the process of photosynthesis requires the reaction of a five-carbon sugar, ribulose-bisphosphate (RuBP), with CO_2 to produce two three-carbon sugars. Because the initial products of photosynthesis are three-carbon sugars, this photosynthetic pathway is known as C_3 photosynthesis. Nearly all trees, woody shrubs, and many forbs and grasses have evolved with the C_3 pathway. The C_3 pathway is not especially efficient at converting CO_2 to sugars because 20–40 percent of the sugars produced from photosynthesis is immediately converted back into CO_2 and respired away.[18] C_3 photosynthesis also produces more water loss from tissues to the atmosphere than other pathways such as the C_4 pathway (not described in detail here) or the CAM pathway. However, the advantage of the C_3 pathway is that it requires less metabolic energy to conduct photosynthesis and therefore works well in relatively cool areas and where soil moisture is consistently available for plant roots to absorb.

Unlike C_3 and C_4 plants, CAM plants close their stomates during the day, when high temperatures and the low humidity of air would otherwise cause large amounts of plant water loss. Instead, CAM plants open their stomates at night so that CO_2 can diffuse into the chlorenchyma cells. Because sunlight is still needed to reduce CO_2 to photosynthetic sugars, the CO_2 that enters the stem at night is converted into an organic compound called malic acid and stored in the large vacuoles of chlorenchyma cells. Once the stem is exposed to sunlight, the malic acid is converted back into CO_2 and reacts with RuBP to produce two three-carbon sugars in the same way C_3 plants convert CO_2 to sugars. As noted above, the disadvantage of the CAM pathway is that it is metabolically expensive and therefore is found primarily in plants that grow in arid, high-sunlight environments and in the leaves of plants in tropical forests. It is worth noting that the CAM pathway has evolved separately many different times and is common in some of the most widespread angiosperm families, including Agavaceae, Bromeliaceae, Crassulaceae, Euphorbiaceae, Liliaceae, Orchidaceae, and, of course, Cactaceae.[19]

Due to the high metabolic cost of CAM photosynthesis, the majority of CAM plants behaves like C_3 plants when given adequate water.[20] In addition to opening their stomates at night, these plants also open their stomates during cool morning hours or in the late afternoon and immediately convert CO_2 to sugars using the C_3 pathway. Many CAM plants are facultative, meaning that they can turn the CAM pathway off completely and use the C_3 pathway exclusively when soil water supplies are high enough to outpace demand. However, CAM photosynthesis in saguaros is not a facultative trait. Saguaros never turn off the CAM pathway, and there is currently no evidence that they turn on the C_3 pathway during the day, even under ideal environmental conditions.[21] One way to better understand photosynthetic activity of saguaros is to continuously measure CO_2 concentrations inside an enclosed chamber with a saguaro plant sealed inside (photo in figure 4.5). Measurements previously taken on young potted saguaros reveal that carbon uptake is primarily confined to periods between midnight and dawn (figure 4.5). Over the remaining periods of the day, net carbon uptake is either nearly zero (around midnight) or below zero, indicating that the plants were *releasing* CO_2 from their tissues. These midday CO_2 releases are common in CAM plants and are caused when malic acid that is converted back to CO_2 during the day escapes the plant before it is fixed into sugars by the photosynthetic enzymes.[22] These CO_2 releases can significantly reduce the overall net carbon uptake over the course of the day.

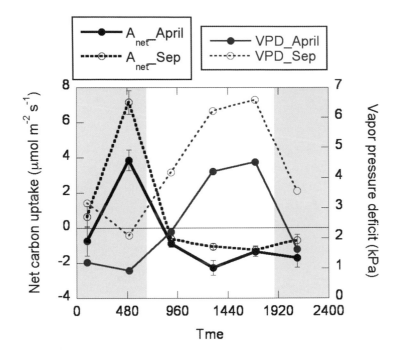

FIGURE 4.5 Patterns of net carbon uptake (A_{net}) measured on eight- to ten-year-old potted saguaro plants. The measurements were conducted in Phoenix, Arizona, using a LiCor 840A CO_2/H_2O gas analyzer (LiCor Inc., Lincoln, Nebraska) connected with flexible tubing to a sealed plexiglass chamber (shown in the photo). The red lines show values of atmospheric vapor pressure deficit (VPD) calculated from air temperature and humidity measurements taken with a nearby micrometeorological station. The error bars represent the standard error of the means.

The photosynthetic activity of saguaro stems not only varies over the course of the day, but also varies seasonally. Seasonal differences are clearly visible in the data shown in figure 4.5, where net carbon uptake was consistently higher over the course of the day in September than in April, even though the plants were watered equally during both periods. At first glance, these seasonal patterns contrast what would be expected, based on the substantially higher vapor pressure deficit of the air (an analog for the evaporative demand experienced by the plant) recorded during the same time as the chamber CO_2 measurements (the red lines in figure 4.5). The higher vapor pressure deficit in September would induce the stomatal guard cells to lose their turgor and reduce the aperture of the stomatal pores relative to April. This simultaneously would reduce water vapor diffusion from the internal tissues to the dry atmosphere and reduce CO_2 diffusion from the atmosphere into the substomatal cavities. However, unlike many columnar cacti species, saguaro growth is primarily confined to the summer and early fall, matching the timing of the summer monsoon. The higher photosynthetic activity (and growth) displayed in summer requires a greater fraction of stored water to be released from the inner cortex to the chlorenchyma tissues to compensate for the higher water loss from the stomatal pores. Thus, saguaros appear to take on a very safe gas exchange strategy in the spring and ramp up their activity in the summer to provide the necessary sugars to support tissue construction.

The "safe" gas exchange strategy of the saguaro is illustrated in depth when compared to other giant cacti species of the Sonoran Desert. Gas exchange measurements conducted before dawn on potted saguaro, cardón sahueso (*Pachycereus pringlei*), senita cactus (*Lophocereus schottii*), and organ pipe cactus (*Stenocereus thurberi*) reveal that saguaros display lower fluxes, including net photosynthesis, transpiration, and stomatal conductance, and higher water use efficiency than the other three species (figure 4.6).[23]

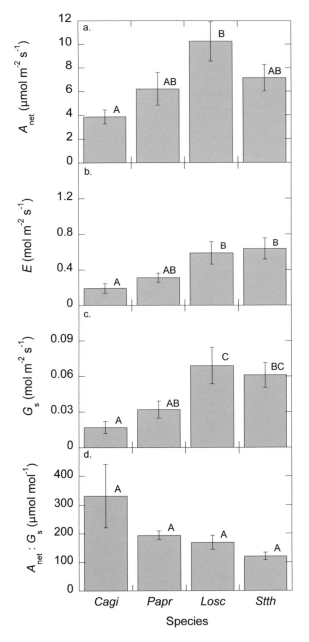

FIGURE 4.6 Whole-plant gas exchange in potted cacti plants measured at predawn in early April 2017. Panel A, mean CO_2 uptake (A_{net}: μmol m^{-2} s^{-1}); panel B, mean transpiration rate (E: mol m^{-2} s^{-1}); panel C, mean stomatal conductance (G_s: mol m^{-2} s^{-1}); and panel D, mean water use efficiency (A_{net} G^{-1}:μmol mol^{-1}) of potted *Carnegiea gigantea* (*Cagi*), *Pachycereus pringlei* (*Papr*), *Lophocereus schottii* (*Losc*) and *Stenocereus thurberi* cacti (n = 5 plants per species). Letters indicate significant differences among species ($P < 0.05$). Error bars represent the standard error of the means.

The combination of high stem volume per surface area, slow growth, and a photosynthetic strategy that limits water loss highlights the extent to which saguaros have evolved to conserve water and tolerate aridity. The form and functional strategy of the saguaro contrast with other giant cacti species like the Sonoran Desert native, senita cactus (*Lophocereus schottii*). An individual senita can have dozens of stems sprouting from its base, with many new stems constructed in a single year (compare the senita with the saguaro in figure 4.4). These stems grow rapidly, but each stem likely only persists for about twenty years, while an individual saguaro can persist for about two hundred years. Unlike the multistemmed senita, the single stem of the saguaro is solely responsible for photosynthesis, resource storage, flower production, and structural support after arms are constructed later in life. Thus, an individual saguaro can only achieve its massive size and long lifespan by taking anatomical and physiological precautions to maintain a functional stem though periods of drought and other challenging environmental conditions.

Five

GENOMICS OF THE SAGUARO

MICHAEL SANDERSON

T he first plant species to have their genomes completely decoded—sequenced—were *Arabidopsis* (in 2000), an inoffensive little annual plant of interest mainly to plant geneticists, and rice (in 2002), which feeds half the world. Since then, as sequencing technology has gotten more sophisticated and less expensive, many other genomes of plant species have been assembled. In the last few years, "charismatic" plants with little economic importance but great potential importance for understanding the evolution of plant diversity have been receiving increased attention. Carnivorous plants, plant parasites, mangroves, succulents, orchids, and, yes, cacti, have garnered the attention of the genomics world.

The Genome(s) of Saguaro

A genome is the sum of all genetic information stored in a nucleic acid called DNA, found as a double helix in cells of almost all life on earth. It controls the vast biochemical machinery of the cell itself, and in multicellular organisms like us and saguaros, it controls interactions between cells and, early in life, the complex growth and development of a macroscopic organism from a single cell. The information in DNA is encoded by four different chemical bases, abbreviated by A, T, G, and C, which are paired with a base on one strand and its biochemically complementary base on the other strand ("base pair," abbreviated bp). A DNA sequence is thus written as a sequence of letters corresponding to the bases on one of these strands: for example, ACGTGCAGCT. . . . The DNA sequence is a blueprint for an intricate process that starts with the transcription of the DNA's information into another kind of nucleic acid called RNA, which acts as a messenger within cells. These messengers in turn are used to guide the assembly of enzymes and other proteins that make cells work.

Small and simple organisms such as bacteria have genomes with a few million base pairs of information. The human genome has about 3 *billion* base pairs. The first sequenced plant genome, *Arabidopsis*, has a genome size of 135 million base pairs (Mbp), but the range of sizes in plants is remarkably large, from the "tiny" 63 Mbp genome of the carnivorous plant *Genlisea* to the preposterously large 150 billion base pair (Gbp) genome of the monocot *Paris japonica*—50 times larger than the human genome! The saguaro genome size has been estimated to be 1.4 Gbp, or about half the size of the human genome. Some cacti have much larger genomes. *Mammillaria* species, a diverse group of small, ball-shaped cacti found mainly in Mexico, and *Weberbauerocereus*, a columnar cactus from Peru and Chile, top out around 7 Gbp. Kew Gardens in London maintains a large database of genome sizes of plants (https://cvalues.science.kew.org).

A plant cell actually contains not one but three genomes, each confined to a particular subcellular compartment, or organelle: either the nucleus, chloroplast, or mitochondrion (figure 5.1). The smallest genome in the plant cell is the chloroplast genome. It is circular, unlike the nuclear genome but like its blue-green algal ancestors, and in most plants is about 120,000–160,000 bp in size. The chloroplast genome has only about one hundred genes, many of which are involved in photosynthesis, which takes place on specialized membranes within the chloroplast. Photosynthesis harvests carbon from atmospheric CO_2, using the energy of sunlight, providing food and energy for the plant.

The saguaro chloroplast genome has been fully sequenced and is quite unusual.[1] At 113,000 bp, it is the smallest chloroplast genome of any flowering plant that still undergoes photosynthesis. There are flowering plants with smaller chloroplast genomes, but these are parasites or carnivores. Saguaro, in its very sunny world, is hardly that,

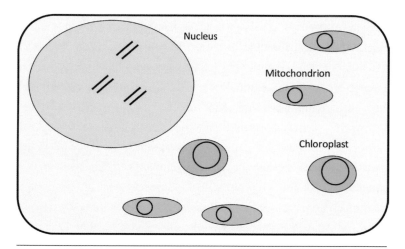

FIGURE 5.1 Genomes of the plant cell. Shown is a schematic diagram of a typical plant cell that engages in photosynthesis. Each of the three cellular compartments, or organelles, has its own genome. The chloroplast (green) and mitochondrial (yellow) genomes evolved from ancient bacteria-like circular genomes and are quite small, compared to the nuclear (gray) genome, which consists (in a diploid plant like saguaro) of pairs of linear subunits called chromosomes. In typical plants the organellar genomes are no more than one hundred thousand to one million bp in size, whereas the nuclear genome may have billions of bp of genetic information.

but it has still undergone an evolutionary reduction in its chloroplast genome. This reduction occurred via two mechanisms: first, most plant chloroplast genomes have a portion of their genome, about 20 kbp, that is exactly duplicated and inverted in its orientation. In saguaro, this large inverted repeat is missing one copy. The functional significance of this is not clear. Second, the saguaro chloroplast genome is missing eleven *ndh* genes found in most other plant chloroplast genomes. These genes all encode proteins that bind together in the NDH complex (an abbreviation for "NADPH dehydrogenase-like complex"), which plays a role in storing energy in the chemical compound NADP during photosynthesis. So, between the loss of these genes and the loss of one copy of the large inverted repeat region, saguaro has a very lean chloroplast genome. Why?

Clues to this might be sought in the rare instances in which other plants have similarly lost some or all of these genes, but these are an odd assortment. Some have lost the ability to photosynthesize and have gone over to a parasitic lifestyle (*Cuscuta, Epifagus*). For them, the plastid genome is often reduced even further than in saguaro. Among plants that still undergo photosynthesis, those that have lost *ndh* genes include some aquatic plants (*Najas*), some carnivorous plants (*Genlisea*), and some that live in extreme hot and dry environments (*Ephedra* and *Welwitschia*, the latter native to the extremely arid Namib Desert of southern Africa). But others that have lost these genes are less exceptional in their lifestyles, such as conifer trees like pines. Clearly, there is no obvious adaptive explanation related to the environment or lifestyle.

However, plastid genomes have a long evolutionary history. Plastids and their genomes originated at the dawn of all complex cellular life by some sort of symbiosis between simple bacteria-like cells, one of which was able to photosynthesize. To survive as a free-living organism, the genome of that ancient photosynthetic cell must have been much larger than a modern chloroplast genome, and their chloroplast genomes have undoubtedly undergone more than a billion years' worth of reduction in size. Most of the genes formerly found in chloroplast's bacterial ancestor have been transferred to the nuclear genome of plants, leaving just the one hundred or so found today.

We can see evidence of these gene transfers almost in real time still. In the nuclear genome of rice, for example, there is an almost full-size copy of its chloroplast genome, and there is evidence of a very high rate of copying of genes from chloroplast to nucleus ongoing—almost all of which leads to degraded and deteriorating nonfunctional copies. In saguaro, too, there is evidence of hundreds of nonfunctional copies of the *ndh* genes in the nuclear genome, as if the transfers took place but did not take. The mystery is how the plant cell can get by without apparently *any* functioning copies of these genes in either the nuclear or chloroplast genomes.

The mitochondrial genome of saguaro has yet to be assembled. In most plants, it is several times larger than the chloroplast genome, has relatively few genes, which evolve very slowly compared to the other two genomes, but has a rapidly evolving *arrangement* of these genes with respect to each other. Mitochondria are the organelles where respiration, which provides energy to cells, occurs in plants, animals, and other complex organisms. It is a profound curiosity that the mitochondrial genome of humans and other animals is very unlike those in plants: animal mitochondrial genomes are tiny, just 16,000 bp or so, with only a dozen genes, compared to plants, which typically have twenty to fifty times larger genomes.

The genome in the nucleus is thousands of times larger than the other two organellar genomes combined, and it contains almost, but not quite, all the essential genetic information in the plant cell. When the phrase *whole* or *complete genome sequence* is used, it typically is referring to the nuclear genome. In land plants like saguaro, the nuclear genome is packaged into linear subunits called chromosomes, and most land plants have these in pairs; hence, they are called *diploids.* Saguaro and most other cacti have eleven pairs of chromosomes. Not infrequently, a new plant species evolves via a process in which these chromosomes are entirely duplicated, leading to a *tetraploid.* A close relative of saguaro, the cardón sahueso cactus of Sonora and Baja California, *Pachycereus pringlei,* is tetraploid, for example.

All modern genome sequencing projects infer or "assemble" a complete genome sequence in much the same way. First, DNA is chemically extracted from tissues. Second, the DNA is broken into billions of small pieces, often only a few hundred base pairs long. Third, these pieces are sequenced from one or both ends for a short distance of perhaps only 100 or 150 base pairs (newer technologies are lengthening this distance dramatically). Fourth, and here is where things get interesting on the computing side, the pieces, or *reads,* are put back together again like a giant jigsaw puzzle.

The scale of this puzzle is remarkable. In the data set we used to assemble the saguaro genome,[2] there were 660 million pairs of reads, most of which were 100 base pairs long (with a small number 300 base pairs long). That adds up to about 140 billion base pairs of sequence reads. This is 100 times larger than the genome size of saguaro, at 1.4 billion base pairs, which is good, because it means a very large number of these reads overlap with one another. In fact, at any given location in the saguaro genome, there are roughly 100 of these reads that overlap, which is what makes the "puzzle" reconstruction feasible.

Ideally, the computer programs that do the genome assembly would be able to reproduce the exact sequence for each of the eleven saguaro chromosomes, but unfortunately this is still not feasible without spending orders of magnitude more time and money. Such chromosome level assemblies are still available for only a limited number of organisms, most of which have genome sizes smaller than saguaro's. The main reason that genome assembly in plants is computationally hard, even when every base in the genome is sampled by 100 reads of sequence data, is that plant genomes are littered with regions of highly repetitive DNA. It is a little like trying to assemble the word "ABRACADABRA" from word fragments that are only two letters long. Notice that "BR" and "RA" and "AB" are all repeated. In plant genomes as large as saguaro, using this many sequence reads of relatively short length, it is just not possible to assemble entire chromosomes when there are many repeats in the genome. Thus, our assembly consists of not 11 but about 57,000 pieces! The largest of these is 649,000 bp long, and there is a highly skewed distribution of sizes: half the pieces are larger than 61,500 bp, and half are smaller. Of the 1.4 billion bp known to be in the genome, the assembly contains about 1 billion. The remainder simply could not be assembled from the short reads obtained during sequencing, probably because of extensive repeat regions.

On the plus side, these pieces on average are large enough to contain a single gene, and the evidence suggests that the saguaro assembly captures 90 percent of the genes likely present in its genome. To sort this out from a new genome assembly, a process of *annotation* is undertaken, an examination of the genome sequence, together with other lines of evidence, to attempt to label various features of the genome. The features include genes and their components, RNAs that act as genes (rather than as messengers, see above), and various other elements in the genome that have functional significance. The evidence used to annotate a genome includes, importantly, annotated genomes from close relatives and computer models of how sequences are organized in genomes (describing the predicted structure of genes, for example).

In the end, we found 28,292 protein coding genes in the saguaro genome, with an average length of 4,800 bp. This is a bit more than the 27,000 found in *Arabidopsis* but fewer than the 36,000 of rice, so there does not seem to be anything particularly exceptional about the overall number of genes. However, these genes, which comprise the bulk of the "information" content of the genome, occupy only 14 percent of the saguaro's whole genome. The largest fraction of the remainder of the genome, some 58 percent, is made up of short, and highly repeated, DNA elements. Most of that consists of fascinating little beasts called transposable elements. Transposable elements, which are "mobile" and literally move around within genomes, were first discovered in maize by the plant geneticist Barbara McClintock, which earned her the Nobel Prize decades after her initial discovery of them in the 1940s. In saguaro, fully 23 percent of the genome is made up of one particular kind of transposable element, the gypsy retrotransposon. In size, structure, and gene content, these retrotransposons are very similar to retroviruses, such as HIV, except they lack the gene for a key viral envelope protein that allows them to exist stably outside of a host cell. Retrotransposons seem to promiscuously multiply and move around genomes but do not escape to become infectious. Indeed, they are most likely the remnants of ancient viral infections. One of the great mysteries of genome evolution is why genomes have not purged themselves of these abundant DNA sequences, which seem to play very little role in the function of the organism but which require extensive cellular resources to duplicate every time the cell divides during its life cycle.

As in other plant genomes, many of these saguaro genes are found in *gene families* of related genes. In saguaro, there are several thousand such families having two or more related genes. These families arise by evolutionary processes that duplicate genes (and sometimes delete them later). Once duplicated, a gene is free to evolve a new function, while its original function is still maintained by the other copy. This is thought to be one of the key ingredients in the long-term evolutionary diversification of genomes. In saguaro, one of the largest families contains more than five hundred genes coding for pentatricopeptide repeat (PPR) containing proteins. These proteins act as regulators of the action (expression) of other genes, primarily genes in the chloroplast and mitochondrial genomes.

A smaller gene family that may play a special role in how saguaro thrives in the hot dry Sonoran Desert is a set of fourteen *pepc* genes that make various versions of the enzyme phosphoenolpyruvate carboxylase (PEPC). Several are involved in photosynthesis in acquiring carbon from atmospheric CO_2. The history of this gene family across plants has been well studied, and it is clearly ancient. Prior even to the origin of land plants, there was a duplication of an ancestral *pepc* gene into two. One of these copies has remained relatively the same ever since, but the other has continued to diversify by subsequent duplications and changes in function. In particular, many succulent plants living in hot, arid regions have a specialized kind of photosynthesis, called CAM ("crassulacean acid metabolism," after one of the plants that does this). Rather than open their stomates, or pores, on their leaves or stems (in the case of many leafless cacti) during the day, so that CO_2 can be drawn in and O_2 given off by typical photosynthesis, CAM plants open their stomates only at night to limit water loss during peak daytime temperatures and low humidity. They then use PEPC to fix the carbon atoms from CO_2 into a storable form, as malic acid, at night only. During the day, when light energy can be harvested for photosynthesis, the carbon is released back into the normal photosynthetic pathways. It has already been established that cacti and their relatives that undergo CAM have more duplicate copies of PEPC that have been recruited over time to be biochemically specialized for CAM pathways. The saguaro genome assembly confirms the exact number of these and permits their evolutionary history of duplication and changes in function to be reconstructed in great detail.

We have already mentioned that some plant species are polyploid, having undergone a process of whole genome doubling. In fact, plants are unusual in that their genomes appear to tolerate this kind of radical structural change relatively well. Some estimates suggest 20–30 percent of living plant species are recent polyploids.[3] Genome doubling has such a radical effect on chromosome organization that new polyploids are usually instantly unable to interbreed with diploid parents from which they come, making them "instant species." While all of these facts about polyploidy in plants have been known since the mid-1900s, something new emerged with the sequencing of the first plant genomes. Close inspection of *Arabidopsis* chromosomes and gene families seemed to hint that its genome may have undergone a duplication of its whole genome ("whole genome duplication," or WGD), perhaps of the type that goes on regularly with modern polyploid plant species but much more ancient. One line of evidence was lengthy tracts of approximately the same DNA sequence found on two different chromosomes, but these were broken up with lots of intervening DNA unique to the different chromosomes. Another line of evidence came from comparisons of pairs of genes in gene families. If a whole genome duplicates, each of its genes is obviously duplicated. Since the WGD occurs at one point in time, the amount of time that each member of a gene pair has diverged is about the same. Since over time genes accumulate mutations, making them progressively more diverged from one another, the average divergence between pairs of genes descended from a WGD should be about the same. By looking at a histogram of these divergences, investigators noticed a spike, or sometimes even multiple spikes, signaling one or more WGDs in the history of *Arabidopsis*, some dating back tens of millions of years. When this analysis is undertaken on saguaro genes, such a spike is also seen. It, too, has undergone a whole genome duplication at some point in the distant past, probably just before cacti originated. This is the same duplication found by other investigators looking at a broad sample of plant families related to cacti. These spikes are not quite the iridium spike marking the cataclysmic meteor impact that coincided with the extinction of dinosaurs 65 million years ago, but they have led to an equally revolutionary upheaval in understanding about major features of plant evolution.

Phylogeny of Cacti and Genomic Clues on the Origins of Saguaro

By comparing the genome of saguaro to those of other cacti, we can improve our understanding of its evolutionary origins and relationships to other cacti. Since Darwin's *Origin of Species*, the word *relationship* in evolutionary biology has come to refer to one specific thing: the relative timing of common ancestry between species. The only illustration in Darwin's book set the stage by showing a branching diagram—*phylogeny*—of evolutionary relationships. In it, a branch point signals two things, a speciation event—that is, one species becoming two—and the most recent common ancestor shared by all the descendants of that branching point in the tree (figure 5.2). Two species that share a common ancestor that is more recent in time than a common ancestor they share with any other species are "more closely related." For example, in this framework birds and crocodiles are more closely related than crocodiles and lizards because birds and crocodiles share a more recent common ancestor—despite some notable similarities between crocodiles and lizards. All the descendants of a single common ancestor are referred to as a *clade*. Many familiar taxonomic groups, like mammals or flowering plants, are clades, but some equally familiar groups are not, like reptiles and bacteria (figure 5.2).

Darwin did not propose any method to infer his branching diagram from data, but nearly a century later other authors did. In fact, a combination of rigorous quantitative methods for phylogenetic inference and exponentially increasing amounts of DNA sequence data led to an explosion of phylogeny reconstruction beginning in the 1980s. All these methods rely on the basic observation that as DNA sequences are passed on from generation to generation and from ancestral species to more recent species, occasional changes that occur in the sequences are shared between groups of related

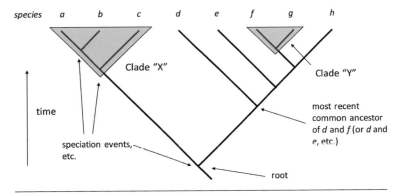

FIGURE 5.2 Phylogenetic trees and some terminology to describe them. This phylogenetic tree shows the evolutionary relationships between eight species labeled *a–h*. Time runs forward from the root of the tree toward the present, and along any path in the tree one can imagine a series of ancestors giving rise to their descendants. Splits in the tree correspond to speciation events where new species are produced (some, but not all, splits are labeled). Each of these splits also marks the location of a "most recent common ancestor" of two or more of the extant (living) species (again, some are labeled as such). It is often useful to describe collections of species on such a tree. The most common term is *clade* or, more pedantically, *monophyletic group*. Two clades are labeled here as *X* and *Y*, but any group comprising one (ancestral) point on the tree together with all of its descendants forms such a clade. Finally, the two clades or species descended from a splitting event in the tree are called *sister or sibling groups*. Thus clade *Y* and species *h* are sister groups. Using these terms, one can verbally describe any phylogenetic tree efficiently and unambiguously.

organisms. Therefore, by comparing sequences of the same gene or genes in different living species, we can infer the historical pattern of common ancestry of those species. Today, this discipline of phylogenetic biology has morphed into the modern hybrid discipline of phylogenomics, which has grown up around the new genome-scale data sets and methods required to handle their data set size. This area has benefitted from interdisciplinary efforts of genomicists, systematists, mathematicians, and computer scientists, and new technologies and algorithms applied to genomic data sets have revealed some surprising biological results, which we see in cacti as well as many other clades.

Historically, the first stabs at understanding the phylogenetic relationships of saguaro among the cacti were very coarse. Saguaro was placed in the genus *Cereus* (as *Cereus giganteus*) by the first scientist to formally describe it, George Engelmann. This genus was destined to be a kind of catch-all group for large columnar cacti of North and South America. However, these landscape-altering, charismatic cacti received a great deal of study, and it quickly became clear that the North and South American columnar cacti in *Cereus* were not closely related. In particular, studies of the flora of Mexico and the southwest United States revealed a large assemblage of columnar cacti with seemingly much closer relationships to each other. By the 1970s, a variety of lines of evidence from comparisons of morphology and biochemistry began to converge on the conclusion that saguaro belongs to a group of cacti referred to at that time as the tribe Pachycereeae. A tribe is a taxonomic rank above genus but below family. Depending on the authority, the number of genera in Pachycereeae varied wildly between 6 and 23 but more or less consistently included the approximately seventy species of columnar cacti in North America. In addition to saguaro (*Carnegiea*), which has but a single species, this included the much larger genera of *Pachycereus*, *Stenocereus*, and *Neobuxbaumia*, some members of which have at various times been put forward as the closest relative of saguaro.

Biologists are interested in cacti, and, not surprisingly, several phylogenetic trees of cacti have been reconstructed over the last twenty years, which included representatives of Pachycereeae, some even with saguaro (figure 5.3). Those using DNA sequence data have relied on one or a few genes from either the chloroplast genome, or a small number of nuclear loci that have been widely used in plant phylogenetic reconstruction. Unfortunately, no study has included

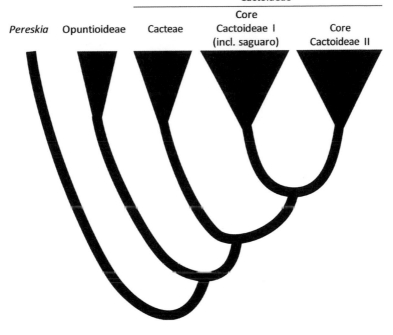

FIGURE 5.3 Phylogenetic tree of all cacti based on molecular studies using DNA sequences from a few genes have revealed the broad outlines of cactus phylogeny, while leaving many details still unresolved. All cacti have specialized arrangements of spines in clusters called areoles. Most species are contained in two large clades. The smaller is the opuntioid clade, including chollas and prickly pears in the genus *Opuntia*, and relatives. The larger by far is the cactoid clade, having some 80 percent of all species of cacti and including not only all the columnar cacti, such as saguaro, but the barrel cacti, hedgehogs, mammillarias, epiphytes, and many others. Opuntioids have ephemeral leaves in their life cycle and possess specialized tiny spines called glochids, which make them exceptionally nasty to herbivores and cactus fanciers alike. Cactoids lack leaves entirely, and photosynthesis has been taken over completely by the stem. They also lack glochids but are very diverse in their overall architecture, ranging from giant columnar forms like saguaro to much more diminutive growth habits. Some enigmatic species of cacti are not found in either of these clades. Of particular significance are the seventeen species in the tropical genus *Pereskia*. Different representatives of this group may trace their ancestry back to near the origin of all cacti, which is consistent with their unusual and "primitive" (for cacti) feature of retaining fully functional photosynthetic leaves.

all the representatives of these genera of Pachycereeae, leaving any comprehensive conclusion about the phylogeny of these plant species somewhat up in the air. Despite this, some consistent conclusions have emerged. In the most comprehensive phylogeny of cacti to date, Hernández-Hernández et al. sampled five plastid and nuclear genes for 224 cacti and inferred a phylogenetic tree that included many Pachycereeae (but far from all).[4] In their tree, the closest relatives of saguaro are species in the genus *Pachycereus*, if we include *Lophocereus schottii* (the senita cactus) as a member of this genus (as it has been in some taxonomic works). Other genera that have been proposed as close relatives of saguaro, particularly *Neobuxbaumia* and *Stenocereus*, were more distantly related to saguaro than *Pachycereus*. Essentially similar results have been obtained from other DNA-based studies that sampled overlapping but different sets of species and genes.

Before delving into the mysteries of where saguaro fits exactly in this broader phylogeny of cacti, it is worth remarking on some relevant evolutionary patterns that appear at the larger scale of all cacti. In particular, the interesting historical mistake of placing saguaro in *Cereus*, mentioned above, has now been placed in a clear evolutionary context. Much of the Hernández-Hernández et al. paper on cactus phylogeny was aimed at understanding the evolution of the remarkable diversity of growth forms in cacti.[5] In addition to the giant columnar cacti, such as saguaros, there are forms that are shaped like barrels, ones that are small and globe-shaped, highly branched and shrubby, vines, ephiphytes, and many others. One thing a phylogeny can reveal is how many times a trait evolves in the evolutionary history of a clade. In cacti, the evidence is clear that the giant columnar form exhibited by saguaro has evolved more than once, perhaps many times. In fact, we now infer that each of the two main clades of cactoid species, core Cactoideae I and II contain one or more origins of this growth form. In core Cactoideae I, in addition to arising at least once in Pachycereeae, large columnar forms are also seen in the genera *Neoraimondia, Corryocactus, Armatocereus,* and *Eulychnia,* which are only distantly related to Pachycereeae. In core Cactoideae II, a number of genera include large columnar species, including *Browningia, Cereus, Echinopsis, Stetsonia,* and others.[6] Here, as in core Cactoideae I, the distribution of species in the phylogeny that are not columnar strongly implies that the giants have evolved several times.

These independent origins are strongly correlated with their present geographic distribution. The columnar cacti of Pachycereeae are centered in Mexico, extending north and south from there in North America with only a few exceptions. The genera *Neoraimondia, Eulychnia,* and other giant columnars in this core Cactoideae I are distributed, by contrast, in western South America in the Andes and along the coast. In core Cactoideae II, *all* the large columnar species

FIGURE 5.4 Convergent evolution of the giant columnar growth form seen in (A) saguaro, and many other North and South American cacti, as in (B) *Echinopsis atacamensis* in Argentina. Photos by Michael Sanderson (A) and username Harvardtl – own work, CC BY 4.0, https://commons.wikimedia.org/w/index.php?curid=66635488 (B).

are found in South America, as are almost all other cacti in this clade. Clearly there have been multiple origins on the two continents.

The term for multiple origins of a trait in distantly related species is *convergent evolution*. Dramatic textbook examples in animals are powered flight and the wings of bats, birds, pterosaurs, and insects. Since Darwin, instances of convergent evolution have been viewed as outstanding examples of the power of natural selection to shape adaptation to extreme environments or to exploit novel "adaptive zones" opened up by key innovations like wings. The saguaro's giant growth form is now a well-supported additional example of convergence. The succulent morphology found in all cacti, but especially emblematic of these giant columnar species, is itself a textbook example of convergent evolution with even more distantly related plants like some giant African species of *Euphorbia*. As yet, any explanations for the convergent evolution in the growth form exhibited by saguaro are speculative, but further studies of its ecology, physiology, and genomics, in comparison to its giant relatives, will shed more light on this question.

The more recent evolutionary history of saguaro, however, clearly involves its relatives in Pachycereeae. Using the genome sequence of saguaro, it was possible to expand the handful of genes used by Hernández-Hernández et al. in their study of all cacti to thousands of genes to reconstruct saguaro's phylogenetic history and origins.[7] However, although increasing the number of genes used to build phylogenies beyond a handful was desirable, it would be overkill to sequence all 28,292 of the saguaro genes in each of its potential relatives. We therefore did much less exhaustive sequencing of three of saguaro's closest relatives and assembled their genomes in a more low-resolution fashion. The three cacti are all columnar cacti in the tribe Pachycereeae. The most similar of these to saguaro is the cardón sahueso (*Pachycereus pringlei*), a giant columnar cactus found in Sonora and Baja California in Mexico. It is among the tallest of all cacti and is more massive than many saguaros, typically with thick branches that emerge closer to the base of the plant. We also sequenced the organ pipe cactus (*Stenocereus thurberi*), which is widespread in northern mainland Mexico and Baja California and has substantial populations in the United States in the vicinity of Organ Pipe National Monument in Arizona. A final columnar sequenced was the senita cactus (*Lophocereus schottii*), also mainly distributed in northern Mexico, with just a few small populations in the Organ Pipe National Monument area. Organ pipe and senita are substantial columnar cacti but smaller than the other two, with shorter life spans. To provide a phylogenetic frame of reference, we also sequenced the genome of a much more distantly related cactus, *Pereskia humboldtii*. This "outgroup" allows us to determine which differences in the genome sequence data among the columnar cacti came first by comparing them to that more distant relative. *Pereskia* is not very cactuslike in many respects. It has leaves and lives in tropical environments, but its flowers, fruits, and spine arrangement are a dead giveaway that it is indeed a cactus.

Despite the much lower resolution genome assemblies constructed for the three close relatives of saguaro and *Pereskia*, it was still possible to find more than four thousand genes (or parts of genes) that were present in the genome assemblies for all five of these cacti. Each of these provides its own independent line of evidence about the phylogenetic relationships of saguaro. Prior to some of the revelations that have emerged in the last few years from studies like this that use whole genomes to reconstruct phylogenies, the standard approach to harnessing all this information would be to simply combine the sequences of the four thousand plus genes into a single giant mega data set, as if it were a single large gene sequenced from all five species. This is still an interesting exercise, and when it is done for these data, a phylogeny emerges that agrees well with previous work based on just a few genes. The closest relative of saguaro among these plants is cardón sahueso, next is senita, then organ pipe (figure 5.5).

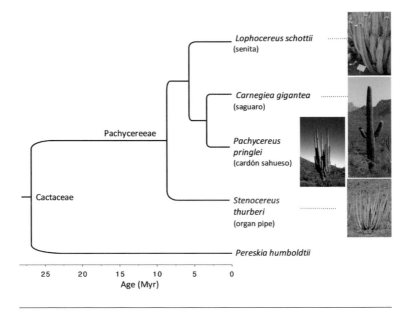

FIGURE 5.5 Phylogenetic relationships of saguaro relative to other Pachycereeae species. Estimates of the age of speciation events can be seen relative to the geologic time scale at bottom. Figure first published in Copetti et al. 2017.

Unfortunately, this is not a complete sample of all species in Pachycereeae or even the genus *Pachycereus*, so although *P. pringlei* is the closest relative to saguaro among the species sampled, there may be other unsampled species that are closer still. One possibility suggested by Hernández-Hernández et al. is *P. pecten-aboriginum*, a columnar cactus broadly distributed from Sonora south along the coast in Mexico to Oaxaca.[8] It was the closest relative of saguaro in their molecular phylogenetic study and together with several other species of *Pachycereus* was closer to saguaro than was senita. Unfortunately, their study did not include *P. pringlei*, so we cannot tell from these two phylogenies which species is actually closer to saguaro. However, we can extract the same genes from the plastid and nuclear genomes of cardón sahueso that were used by those authors and rerun their analysis. Based on the fragment of the nuclear *pepc* gene used in that paper, *P. pringlei* and *P. pecten-aboriginum* (and *P. marginatus* too) are all equally distant from saguaro, forming a clade of three *Pachycereus* species.

Before taking this into the era of phylo*genomics*, we can also make some educated guesses about the geological timing of the diversification of Pachycereeae based on the rates of evolution of these genes. This requires making some strong assumptions about the rate of genome evolution. Such work appeals to a theory about molecular evolution known as the "molecular clock," which has been simultaneously controversial and irresistible to evolutionary biologists. In a nutshell, the molecular clock supposes that the rate of evolution of at least parts of genes is constant across different species, and therefore the number of DNA differences between two species' genomes is expected to be twice as much if they split from each other twice as far back in the past. To the extent this is true, the clock can then be combined with fossil evidence from one part of a phylogeny to make predictions about the timing of splits in another part, where no fossils are present.

The conundrum is that we wish to use molecular clocks in cacti in the first place because they have such a paltry fossil record, but we need somehow to have at least one "calibrated" part of the phylogeny of cacti to use the clock at all—that is, one part having a known date. To the rescue are broader phylogenetic analyses across much deeper parts of the phylogeny of plants, including parts that have good fossil records. From these studies, other workers have computed an age for all cacti at about 26 million years ago (MYA). So, we can use that calibration for all cacti, in conjunction with the observed sequence divergence seen between the Pachycereeae species and the more distant *Pereskia*. These lead directly to the inferred dates of figure 5.5, where the split between saguaro and cardón is at about 3.4 MYA, and the next splits are then at 5.7 and 8.7 MYA. The entire Pachycereeae probably dates back to nearly 9 MYA, about a third of the time back to the origin of cacti.

But back to the phylogenetic relationships of our cacti. What insights can be gleaned from having four thousand plus genes in all five cacti compared to the half dozen or so genes used in previous phylogenetic studies of Pachycereeae and other cacti? With this much data, it has proven possible to infer separately a phylogenetic tree for each gene, so rather than building one *species tree* from all the genes combined, we can build a *gene tree* for each gene, and then use these in some fashion to understand the history of the whole genome of the species. One of the surprising revelations to emerge from phylogenomics, the marriage of phylogenetic techniques and genomic data, is that not only do some of these gene trees disagree with each other, but in some cases, *many* of them do. In phylogenomic studies of primates, mouse, and *Drosophila*, undertaken shortly after those genomes became available, it quickly became clear the disagreement between these gene trees was more the rule than the exception. Since then, studies in plants like rice and tomato and their wild relatives have shown the same pattern.

At first it may seem surprising that different genes in the same set of species can have different phylogenies, but this can happen for several reasons that are very relevant to the particular biology, ecology, and history of saguaro. First, from basic genetics we know that different genes in a genome can lead semiautonomous lives: they may be on different chromosomes, in which case they assort independently when DNA is replicated during meiosis, or even if on the same chromosome they may be far enough apart that a genetic process of crossing over or recombination can make them assort independently from one another as if they were on separate chromosomes. (This contrasts with the one hundred plus genes in the plastid genome, which do not undergo recombination, and therefore the gene trees of all genes must be the same.) Then, if there are forces that would allow different genes to have different histories, their genetics at least does not prevent it.

There are two key processes that cause different genes to have different histories. The first is known as "incomplete lineage sorting," an opaque bit of jargon, which obscures what is actually a fairly simple process (figure 5.6). Every species, including the species ancestral to these cacti, is really a population of many individual organisms. At some point, the ancestral species splits to form two descendant species, and the population of individuals likewise splits into two populations. If we sample one individual from each of several living species and sequence some gene, we would expect that the gene tree inferred would match the species tree (figure 5.6A), because, after all, the genes are trapped within the organisms, which are trapped within the populations/species. However, the history of the gene does not have to track the species exactly, precisely because the species is made up of many individuals (figure 5.6B and 5.6C). The phylogeny of a gene starting out in a single individual organism in a population splits when it is passed on to two children, but this may happen long before the population splits into two species. It is possible that the history of that particular gene, if both copies survive to appear in the modern species, may disagree with the species history (figure 5.6D). The two factors that make this disagreement more likely are large population sizes, because it takes longer for the genes to spread through such populations and a shorter amount of time between the species splitting events.

The second process that can cause gene trees to disagree with each other and with the species tree is "gene flow." Gene flow includes several processes such as hybridization that "move" genes from one species to another. Many species are unable to interbreed with each other—in fact, this is the conventional definition of biological species—but many exceptions exist. Interbreeding can occur during the late stages of the speciation process before a species is completely

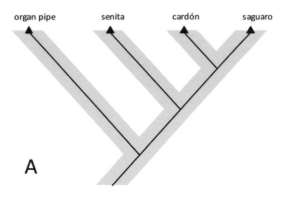

A

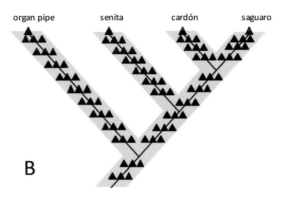

B

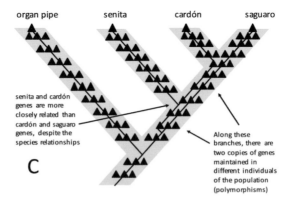

senita and cardón genes are more closely related than cardón and saguaro genes, despite the species relationships

Along these branches, there are two copies of genes maintained in different individuals of the population (polymorphisms)

C

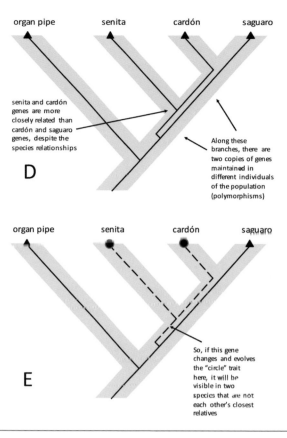

FIGURE 5.6 How gene trees can be different from a species tree. (A) The simplified model of phylogenetic history in which a gene (black line) is contained within a species (gray background). (B) A more accurate model of a gene's history within a species showing its descent through individuals (black triangles) in a population. This is only one of many treelike paths possible in the history of the three genes sampled in the modern species. (C) In this population genetic setting, a gene tree can be different from its species tree. Here the historical split in the ancestry of a gene within one population occurs much earlier than the speciation events, and both copies (alleles) persist while the species undergoes two subsequent speciation events. (D) Here we have simplified panel C by removing the symbols for individuals, making the gene tree relationships clearer. (E) Implications of gene tree–species tree disagreement. If one of the alleles has a mutation that controls an observable trait, when the gene tree is different from the species tree, this observable trait has an unexpected distribution among the species. Here the gene is mutated and a "circle" trait is observed in any species containing the descendant gene with this mutation.

"reproductively isolated" from others. Plants have a reputation for being able to tolerate interbreeding among closely related species and even sometimes between more distantly related species. Cacti, in particular, have this reputation. It is worth noting, however, that just because one can force two species of plants in a greenhouse to interbreed does not mean they ever do so in nature. Many mechanisms, like different flowering times or simple geographic isolation, may prevent it. Interbreeding between species can lead to different gene trees because unlinked genes can take alternate paths through the ancestral species to arrive at the same descendant (figure 5.6).

Both of these processes are expected to cause gene trees to disagree with one another in many plant groups. However, the extent of this disagreement in saguaro and its relatives is still remarkably high. More than one-third of the four thousand plus gene trees disagree with one or more relationships shown in the species tree in figure 5.5. Instead of the simple picture of genome evolution implied by that phylogeny, the real picture of genome evolution in these cacti is a much more complex mosaic of different gene trees. So, if one is interested in the evolution of some gene found in saguaro, say one of the *pepc* genes mentioned earlier, chances are good that its closest relative is not found in cardón sahueso but rather in senita or organ pipe, despite the fact that cardón sahueso is more closely related to saguaro.

It is not easy to sort out how much of this disagreement is due to incomplete lineage sorting versus gene flow. There is clear evidence of both happening in Pachycereeae. Estimates of the time along branches and ancestral population sizes are consistent with incomplete lineage sorting being the primary source of gene tree conflict. This is particularly acute in these cacti because they are so long lived. Saguaros may take thirty-five to seventy-five years before they flower and reproduce, which makes the number of generations between speciation events in them smaller than it would be for shorter lived plants (which accentuates incomplete lineage sorting). However, there are asymmetries in the distribution of all the conflicting gene trees that can only be explained by gene flow between some but not all of these cacti following speciation. The various historical scenarios are only weakly distinguishable from one another based on even these genome scale data, but they consistently point to gene flow into cardón from either organ pipe or some other species not sampled (or extinct).

Several cases of natural hybrids between genera are known among cacti. One of these may be relevant to this history of gene flow in Pachycereeae. In central Baja California, there is a small region near El Rosario where the geographic range of cardón sahueso overlaps with another columnar cactus species from Pachycereeae (not sampled in our study), *Bergerocactus emoryi*, which is a much smaller species. In the late 1800s, the naturalist Charles Orcutt discovered a few individuals having features intermediate between *P. pringlei* and *B. emoryi* in the same location, but very few individuals were discovered over the next half century. Eventually these were described as a new species, *Cereus orcuttii*, then renamed *Pachycereus orcuttii*, in line with emerging ideas about the untenable status of *Cereus*. Finally, in 1962, Reid Moran resolved the mysterious status of this plant by extensive comparisons of morphology and analyses of progeny of the plants, showing that *P. orcuttii* was actually an intergeneric hybrid between *P. pringlei* and *Bergerocactus*.[9] Whether *Bergerocactus* is involved in the gene flow seen in the Pachycereeae genomes will only be resolved with further data, but it is a tantalizing possibility.

The muddled phylogenetic history of Pachycereeae genomes might actually explain some of the difficulties we humans have in telling how cacti are related to each other. To the extent that traits observed in the morphology, biochemistry, or other characteristics of cacti are determined by genes whose histories are diverse and different from the true species tree, then these traits will also be distributed among cactus species in a way that does not nicely track the simple species tree in figure 5.5.

The Taxonomy of Pachycereeae and Saguaro

Although genomic evidence solidifies our understanding of the relationships of saguaro within Pachycereae, we do not yet have a sufficiently complete sample of species to think seriously about reclassifying them. The key issue with saguaro has always been "to what genus does it belong?" Partly this is because currently the genus is "monotypic," meaning that it contains only one species. A monotypic genus is a bit agnostic since it tells us nothing about its relatives. If anything, it merely suggests that it is distantly related to species in unspecified other genera. However, a recent study showed that less than 50 percent of *non*monotypic genera of cacti were clades,[10] because they are missing some of the descendants from their most recent common ancestry (figure 5.2), so perhaps falling back on monotypic genera is "better safe than sorry." The concern with saguaro is that *Pachycereus*, which has, from time to time, been proposed as a better home genus for saguaro, might not be monophyletic. For example, saguaro might be more closely related to some *Pachycereus* species than some *Pachycereus* are to other *Pachycereus* species. Indeed, this is the picture suggested by Hernández-Hernández et al., albeit based on only a small number of genes.[11]

Taxonomists often pull out or "segregate" species they regard as highly distinct from their relatives, and modern phylogenetic classifications have tried to undo this to some extent to reemphasize common ancestry rather than outward appearance. Thus, if the Hernández-Hernández et al. phylogeny is proven to be correct,[12] the species of *Pachycereus* in it do not form a monophyletic group. *Lophocereus schottii* and saguaro from our genome studies are both nested within a clade that contains all of these species. One can adhere to the goal of having classification reflect phylogeny with two solutions, both of which cause name changes. The first would lump both senita and saguaro into a much-expanded genus containing all of these plus

Pachycereus. According to the ancient rules of botanical nomenclature, these would all then become part of a greatly expanded genus *Carnegiea*, because the name *Carnegiea* was published first in the literature. This would disrupt the naming of the many Mexican species currently in *Pachycereus*. The other option would be to break up *Pachycereus* into several new smaller monophyletic or monotypic genera. This would also be disruptive, and therefore neither option is tenable until further sampling and phylogenomics is undertaken.

Genomic Variation Across the Range of Saguaro

Although saguaros are endemic to the Sonoran Desert, their geographic range extends more than 1,000 km from northwestern Arizona to near the southern boundary of Sonora in Mexico. Across this area, saguaros are subject to hot and dry conditions, but these vary in detail. To the northwest, summer monsoon rains constitute a smaller component of their annual water budget than in the far southeast. In Arizona, plants to the far west receive one-third the rain in a year (less than 100 mm near Yuma) than those to the east (300 mm around Tucson). The biological environment also varies across the range of saguaro, including differences in the species and abundances of pollinators, such as bats and birds, and in the plants in the local floras that may provide competition or benefits (e.g., nurse plants) to saguaros. Thus, there are many reasons to expect saguaros to exhibit genetic and genomic diversity across their range as adaptations to these factors.

One widely used measure of genetic diversity at the scale of the entire genome is the nucleotide diversity, π. This number is the probability that two individuals randomly sampled from a population will have two different bases at any given site in their genome. An estimate from 20 individuals across the range of saguaro is $\pi = 0.0024$. This means that any pair of saguaros is expected to differ at one out of about every 400 nucleotide positions in the genome, which amounts to about 3.4 million sites in its 1.4 Gbp genome. These sites are called "single nucleotide polymorphisms" or SNPs (pronounced "snips"). This genetic diversity is low compared to most plants, except for some other long-lived perennials like spruce trees, but similar to the genetic diversity in humans.[13] As in humans, since most of the genome does not consist of genes, most of this observed genetic diversity is in noncoding and possibly nonfunctional regions of the genome, but this still leaves hundreds of thousands of differences within genes between individual saguaros, more than enough to code for adaptive differences allowing populations to survive in different environmental conditions.

Preliminary studies of genomes of individuals sampled in sixteen plants in eight populations across the range of saguaro reveal some additional information about how this genetic diversity is structured. Based on a sample of twenty thousand SNPs across the genomes of these sixteen plants, two patterns emerge. First, populations of saguaros are much more closely related to each other than any is to the closest related species we sequenced, the cardón sahueso. A pair of sequences from saguaro and cardón differ at about 2 percent of their sites, compared to the 0.24 percent of that seen *within* saguaro. It is not surprising that the number of within-species differences is less than between-species differences, but there are many cases where the relative difference is not this dramatic. Assuming that saguaro diverged from cardón sahueso three million years in the past (see above), then the most recent common ancestor of the eight populations sampled must be far younger than this, on the order perhaps of 10 percent or less, so no more than a few hundred thousand years.

The second pattern that emerges is subtle but well supported by the data. It appears that the various populations of saguaro have not differentiated from each other much, but there *is* a consistent difference between the two populations in the far south near the border of

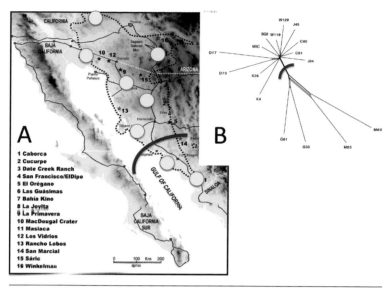

FIGURE 5.7 Population genomics of saguaro. (A) Locations of eight populations across the Sonoran Desert, from which two individuals in each were sequenced. (B) Phylogenetic history of sixteen individuals in these eight populations. Those labeled G ## and M ## are from Guásimas and Masiaca, respectively, in far southern Sonora. The red arc is the position of a strongly supported branch in the phylogenetic tree, seeming to separate these southern populations from the northern ones. Alternative phylogenetic histories supported by different sets of SNPs are indicated by thin rectangles in some places on the phylogeny.

Sonora and Sinaloa, and the remaining saguaro populations to the north (figure 5.7). Recall that gene trees can be different from species trees. Within a single species, all else being equal, we expect gene trees to reflect more or less random mating of individuals across a species range. That means one gene tree of the different individuals will be completely different from another unlinked gene. This is nearly but not quite what we see across gene trees within saguaro. There is an excess of gene trees that have the two southernmost populations closely related. This can happen for a number of reasons, but it most likely reflects at least in part a historical signature of geographic isolation of the south from the north. Much work remains to be done to connect the observed genetic diversity in saguaro to differences in its ecology and local environments. Many methods have been proposed to test statistically for these connections, and having an entire genome's worth of information for many individual saguaros promises to boost the sample size dramatically and make these tests much more powerful than they have been in the past.

Conclusion

The genome of saguaro is large and complex but is beginning to reveal information both about how this charismatic plant makes a living in the harsh environment of the Sonoran Desert and how its ancestry unfolded over the last several million years. In the near future, further sampling of genomes of cacti will almost surely resolve many of the questions surrounding the evolutionary origin and diversification of the saguaro, because the tools of phylogenomics are well enough developed that they can exploit the masses of "big data" now becoming available. Understanding in detail the connection between the saguaro's genome and its physiology, biochemistry, morphology, and life history, however, may well take more time, as many of these questions remain unanswered for even the best understood plant model systems such as *Arabidopsis* and rice. However, studies of these exceptional plants, perhaps in comparison to other plants in equally extreme habitats that have converged on similar adaptations may provide a complementary route to understanding.

Six

THE ANNUAL SAGUARO HARVEST AND CROP CYCLE OF THE PAPAGO, WITH REFERENCE TO ECOLOGY AND SYMBOLISM

FRANK S. CROSSWHITE

Foreword

Frank Crosswhite and I attended Prescott High School in Arizona at the same time. We were both caught up in the science craze in the United States that followed the launching of Sputnik I by the Soviet Union in 1957. Sam, as we all called him, was widely known for his brilliance, his engaging wit, and his wry smile. Following graduation, he and I lost touch for a few decades until we reestablished contact while he was curator of botany at the Boyce Thompson Southwest Arboretum near Superior, Arizona, and I was a researcher at the University of Arizona. It was at Boyce Thompson that Dr. Crosswhite founded the journal *Desert Plants* and established a reputation as a highly skilled botanist, ethnobotanist, and geologist.

The following is an excerpt of an article that appeared in *Desert Plants* in 1980 during the second year of the journal's publication. The depth of Frank's research and knowledge shines through his writing, as well as his deep respect for the O'odham people. He spent countless hours studying their ancient relationship with the saguaro, and this is the product of his research. His decision to include photos by noted photo anthropologist Helga Teiwes was an inspired one.

Nearly forty years have passed since the article was published, and it remains the definitive work on the ethnobotany of the saguaro. During that time the O'odham have rejected the name *Papago*, a label applied to them by Spaniards, replacing it with their own name for themselves, *Tohono O'odham*—Desert People. While Crosswhite acknowledged the O'odham's preference for their own term for themselves, he used the more popular name throughout the article, a practice that since that time has fallen into disfavor.

In compiling this volume on the saguaro cactus, we, the authors, discussed the possibility of composing a more modern or updated discussion of the ethnobotany of the saguaro. After long consideration, however, we concurred that it would be impossible to improve on Dr. Crosswhite's treatment of the subject. His presentation of the O'odham ethnobotany of the saguaro, as best understood through the harvest and crop cycle, is a model for the study of the cultural importance of plants. The saguaro is, for all practical purposes, the O'odham's plant. Different native peoples—Akimel O'odham, Colorado River peoples, Maricopas, Seris (Comcáac), and others, found the plant useful. For the Tohono O'odham, however, it is of such fundamental importance that it is hard to envision them as a people apart from the saguaro. And the annual saguaro harvest represents the deepest connection of this desert people to the plant and to their ancestral lands. Reviewing Crosswhite's descriptions, we could think of nothing we might add to what he wrote. His research and the irreplaceable photos by Helga Teiwes combine to make this document of immense historical and anthropological importance. We are delighted to present excerpts as part of the natural history of the saguaro.

For a comprehensive explanation of the terminology surrounding the saguaro and place names that include saguaros, we refer the reader to O'odham *Place Names* by Harry Winters.

—*David Yetman*

The Calendar of the O'odham in Relation to the Saguaro, Crops, Wild Plants, and the Environment

The Papago have been an indigenous people attuned to the environment of the Sonoran Desert. They have lived in harmony with the *hahshani* (saguaro) and with their environment for hundreds of years. They could almost be considered as much a part of the desert as the cactus itself. They attuned their calendar to the phenology of the saguaro and to the natural progression of the seasons rather than to the sun and its equinoxes. This is at variance with the sun-regulated calendar of the citified Aztecs even though the Papago speak a language which relates them, although distantly, to those people. One wonders how the Hohokam, prehistoric riverine city dwellers of southern Arizona, from whom the Papago have sometimes been claimed to have been descended, regulated their calendar.

The phenological calendar would theoretically be quite effective in regulating the activities of peaceful and democratic agriculturists. Students of Papago ethnology have mentioned the egalitarian principles that have run through the fabric of Papago life before these people fell under the influence of Anglo capitalism, the five-plus-two-day week, and the profit motive. There was apparently little division of labor other than between man and woman. A family unit was relatively self-sufficient in that its dependence was more directly upon the ecology of nature and the agricultural fields rather than on fixed groups of other people.

This is a common peaceful agrarian pattern in many parts of the world where families are dispersed over the countryside as opposed to the "civilized" trend toward living in ever larger cities (partly for protection from without) with rigid division of labor and forced specialization—the butcher, the baker, the candlestick maker, and so forth. Did some "Pax Aztecum" once extend to the countryside of the Papago? The way of life closely attuned to the environment, whenever and however established, endured successive threats of Spanish conquest, Apache warfare, and Mexicanization, less well enduring the ultimate threat—Anglo industrial culture. Once attuned to the environment, the people had to adapt to both subtle and major changes in it. A good example of change is seen in the River Pima, where one of Russell's (1975, 36) informants stated when recounting the names of the Piman months that the year began with "Saguaro Harvest Month," while another informant already gave "Wheat Harvest Month" as the beginning of the year. Wheat had been unknown to the O'odham until it was introduced to them by Father Kino, the Jesuit missionary. The months of the calendar of the O'odham are listed below, summarized from Lumholtz (1971, 76), Russell (1975, 36), Underhill (1939, 124–25) and Saxton and Saxton (1969, 178–80).

1. HAHSHANI MASHAD (June)—the "saguaro [harvest] month," sometimes referred to as the "hot month." This is the time to set up the cactus camp and begin to harvest, process, and eat the fruit—a joyous time but one with hard work.

2. FUKIABIG MASHAD (July)—the "rainy month." Now it is time to finish the saguaro harvest, to start the rain by means of a *nawait* ritual, and then to plant the seeds of beans, corn, and cucurbits.

3. SHOPOL ESHABIG MASHAD (August)—the "short planting month." This is the last chance of the year to get crops

into the ground. If the *nawait* ritual at one village has already started the rains, now other villages should have ceremonies to ask for the rains to continue.

4. WASHAI GAK MASHAD (September)—the "dry grass month." When the rains stop, the desert grass turns brown and the late crops ripen.

5. WI'IHANIG MASHAD (October)—the "month of persisting [vegetation]." Certain food plants characteristically surviving seasonal drought and the onset of cold weather are harvested now. It is autumn, and one might expect to see frost on the pumpkin soon. This has also been called the month of winds, light rain, light frost, or "when cold touches mildly."

6. KEHG S-HEHPIJIG MASHAD (November)—the "month when it is really getting nice and cold." Also translated as the fair cold, pleasant cold, or low cold month. This would be a good time to go hunting.

7. EDA WA'UGAD MASHAD (December)—the "inner bone [backbone of winter] month." This is the dead of winter. It is also referred to as the month of great cold or the month when leaves (for example, of mesquite) fall.

8. GI'IHODAG MASHAD; UHWALIG MASHAD (January to early February)—the time when animals "have lost their fat," then "go into heat" and mate. There is not a lot of activity in the desert, and time seems to hang heavy. This was once a good time for Desert People to go south to work in Mexico or travel north to work among the River People.

9. KOHMAGI MASHAD (February)—the "gray month." This is the time at which the landscape is at a bleak climax of gray. The trees are without leaves, but the flowers are already coming on the cottonwood trees.

10. CHEHDAGI MASHAD (March)—the "green month." Leaves are finally coming on the cottonwood and mesquite trees,

and there are abundant green growing herbs and grasses. A
saguaro ritual is held.

11. OAM MASHAD (April)—the "yellow-orange month." It
is spring, and the beautiful yellow and orange flowers of desert
poppy, brittlebush, desert marigold, and other wildflowers make
the desert colorful and happy, but food stored by humans is
beginning to run low.

12. KAI CHUKALIG MASHAD (May)—the month when
saguaro "seeds are turning black" in the developing fruit. This is
the optimistic name; the pessimistic name was "painful month."
The flowers of spring have disappeared. Hunger pangs were once
a real possibility. A parent might have chosen painful sacrifice
to make sure the children were well nourished. After Father
Kino introduced wheat, the River People harvested grain during
this month, and it became a good time for the Desert People to
demonstrate friendship by helping them with harvesting and
singing.

To Insure a Good Saguaro Harvest

Lumholtz (1971, 60–61) uncovered a Papago ceremony in the month
of March, which dealt with the saguaro. This was the "green month,"
Chehdagi Mashad, the beginning of spring and universally a time of
renewal, including in other cultures, for example, the resurrection
and Easter. "It should be noted that in March, about the time of the
equinox, a ceremony, accompanied by singing, is performed to insure
a good saguaro harvest."

Saguaro seeds were ground and put into a basket with four pieces
of rib from a saguaro skeleton, one at each cardinal point. "Sitting
around the basket, people spend a night singing, while the medicine-
man makes prognostications for the coming harvest." The people who

were present ate the seeds. The pieces of saguaro skeleton were given to four persons, each of whom later, during the saguaro harvest, left one at the base of a separate saguaro. Whether the saguaro ribs were left on the surface or were buried is not stated.

That vitality of saguaro might be enhanced by the sacrificial offering of consecrated saguaro ribs could tie in with a Papago resurrection legend involving life after death: A child sank into the ground and its

FIGURE 6.1 A cactus camp west of the western unit of Saguaro National Monument. The ramada at right makes structural use of a tree of ironwood (*Olneya tesota*). The flat top of an old cookstove is placed on a washtub, which has had the side cut out to make a camp stove. June 30, 1970. Arizona State Museum photograph by Helga Teiwes.

mother asked Coyote to help dig it out. Coyote did this but secretly ate the child and gave the bones to the mother, saying, "Someone must have eaten your son. This is all I could find." The mother then asked Coyote to bury the bones, which he did. In four days something green came up on that spot and in four more days it was a baby saguaro—the first (Wright 1929, 115).

Further renewal rituals, in April and May, which "anticipate the rainy season," were discussed by Underhill et al. (1979, 141–44). Upon renewing ocotillo branches at the children's shrine west of Santa Rosa, ritual speeches dealing with crops and wild food plants were made.

At Mesquite Root village a sacred bundle was opened and "scraping stick" songs were sung "at intervals through the spring as the cactus picking season drew nearer." Reportedly this was done (1) first when saguaro came into bud, (2) next when it came into flower, (3) then when it had its first fruit, and (4) finally just before harvesting.

The Cactus Camp

In the old traditional way of doing things, a Papago family had a winter home, a summer home, and a cactus camp. Papagueria consisted of broad valleys filled with alluvial soil and bounded by small steep mountains (Bryan 1925, 1). In winter, water was reliably found only in such hills so the winter home, called "the well," was established there (Castetter and Bell 1937, 13). Summer thunderstorms brought moisture to the valleys so they could be inhabited and farmed. The summer home was known as "the fields." As part of the annual process of moving from the winter well site to the summer fields, the Papago camped in large groves of saguaro cacti for a few weeks to harvest and prepare the fruit. In more recent years, after the cycle of movement from winter home to summer home had been broken for most families, it remained popular for many years to go out to the cactus camp, harvest and process some fruit, and generally have a good time in the fresh air. But these camps are being used less frequently as time passes. Fisher (1977, 2) noted that there are "fewer jars of syrup or jam found on the shelves in the relatively modern reservation homes" of the Papago. Mrs. Kate Johnson, a Pima from Blackwater, explained to Mahoney (1956, 14–16) that only a few of her people still harvested saguaro fruit because prepared jam could be bought in the store.

Moving to the Cactus Camp

In olden days the families migrated from their winter quarters to the saguaro cactus camp by foot, laboriously carrying baskets, ollas, various needed possessions, and water for rather long distances. Later the trek was accomplished by means of horse and wagon and eventually by pickup truck. Wagons and trucks allowed transport of material possessions and large barrels of water with much less work.

Wickham (1971, 3) wrote that at saguaro harvest time years ago, the Papago would "pack up enough beans and dried jackrabbit for a week, fill clay ollas with water for stewing the fruit, and trek all day to their hereditary picking camp in the rocky foothills where the saguaro flourish." In Wickham's 1971 trip with the Papago, however, the Indians "stopped at the trading post for spaghetti and cold cuts and drove 45 minutes to the Saguaro National Monument, where the federal government permits the Papagos to pick."

When Bowen (1939, 4) wrote of the saguaro harvest, many of the Papago living near Sells were still in the habit of using cactus camps through the season, whereas those living at San Xavier and near Tucson preferred to drive by automobile "to harvest what they can in a few hours and return with it to their homes for the cooking."

Underhill (1951, 44) described arrival in cactus camp: "The girls run to dig up the grinding stones. Mother goes to look at the long poles she uses to hook down the cactus fruit from the top. . . . Many people have passed by this place . . . and they have all seen the poles, but of course they have not touched them . . . everything is safe."

In pre–horse and wagon days, the large cooking ollas represented a real transportation problem. According to Fisher (1977, 2), the Papago placed the huge ollas "in pits and covered them with branches of mesquite and greasewood, before the hole was covered with dirt. And there they would remain until the next season." Densmore (1921, 151) reported that a few ollas "were usually left in the camp and were not disturbed by travellers." Keasey (1975, 29) reported the custom of a Papago lady to leave her fruit harvesting pole from year to year in an upright position securely fastened within the branches of an ironwood tree.

A Camp Ritual

According to Underhill et al. (1979, 21), when a family arrived at its camp there was a customary ritual. "Each person is supposed to take the first ripe fruit that he encounters, open it, extract some of its juicy, red pulp, apply it to his heart, and say a prayer of thanks for having lived another year. Then there would be one, two or three weeks of picking, eating, cooking, drying and storing" the fruit. Applying juice from the first cactus fruit of the season to the body was apparently not unique with the Papago, as Felger and Moser (1974, 259) have reported that the Seri practiced such a tradition, but in this case they applied pulp to "the cheeks and the tip of the nose to bring good luck."

Descriptions of Cactus Camps

Castetter and Underhill (1935, 20) wrote that each family had a camp on some slope where saguaro grew, and there they had a "rough shelter, with a water jar, perhaps a metate, and a cooking pot." Lumholtz (1971, 77) described a cactus camp as consisting "simply of a roof of branches resting on four poles." Keasey (1975, 26) characterized cactus camp ramadas as being "constructed of palo verde poles with a roof made of ocotillo branches, saguaro ribs, palm fronds and tar paper." Densmore (1921, 151) described a thatch in a tree in the cactus camp used for drying the fruit products.

Nabhan (1982) described a recent cactus camp as having a ramada of mesquite poles and plywood patchwork, with furniture consisting of "a card table, three foldups, the bulging single bed and an ice chest. . . . A fruit crate was hung from one of the upright posts, serving as a shrine with a metal crucifix, a porcelain Virgin Mary. . . . On a second post, a cross made of cottonwood leaves was hung to keep the lightning away from camp."

Referring again to Wickham's trip with Papago family to pick saguaro fruit in 1971 (1971, 6), the modern camp was "blanket rolls, pots, groceries, and water jugs" piled under a paloverde tree. The family had to eat their spaghetti cold by the light of a flashlight because campfires were prohibited in the national monument except in designated recreational areas. No shelter was built because the man of the family had to be back to work in his village on Monday, and the cactus camp could only be a weekend affair.

Jollity and the Promise of Good Food and Drink

LaBarre (1938, 232) said that when the saguaro fruits were ripening, the Indians pointed at them laughing and said "see the liquor growing" and sang songs about it. According to Castetter and Underhill (1935, 22), cactus camp was "always an occasion of jollity." It meant the first fresh food, the first taste of something sweet, and was "principally the preparation for the great drinking ceremony which is regarded as responsible for bringing the rains." Underhill (1979, 19) noted that in cactus camp many songs "forbidden up to now" during the year were finally sung. Some apparently were being rehearsed for the coming wine ceremony. Herbert (1969, 3) put it well when he wrote, "In addition to providing a good excuse for a family camp-out, gathering Saguaro fruit has a threefold objective: jam, seeds for meal or chicken feed, and wine for the Rain Making Ceremony." And these were only the chief uses for the fruit; in addition, there was the promise of syrup, dehydrated pulp, oil, snack foods, soft drinks, and even vinegar.

Apparently, it was not uncommon to set camp up a little before the fruit was actually ready. Herbert (1969, 4) observed that "with camp in order and equipment ready, it's time to wait and visit with neighboring families in their camps. Saguaro harvest time is also a social gathering time when politics and weather are discussed along with local gossip." In Wickham's (1971, 6) experience with the Papago harvesting fruit, "you eat the perfect ones on the spot, turning your back to the other harvesters. You chew the oily seeds into a thick syrup before you swallow. It feels luscious and wicked, but no one says anything back at camp. Everyone's mouth is pink and sticky." Niethammer (1974, 23) said that each fruit contained about thirty-four calories and recounted a story written by a priest in the seventeenth century to the effect that after the Indians had been eating the fruit for about three weeks, the priest could not recognize many of his friends because they were temporarily so corpulent.

Tenancy Details of the Camps

Castetter and Underhill (1935, 20) stated that the picking season lasted for about two weeks. Later, Underhill (1946, 41) wrote that for three weeks prior to the wine ceremony the saguaro fruit was "ripe on all the southern hill slopes, and every family left its winter quarters and went to camp in some grove to which it had a more or less hereditary right." Waddell (1973, 217) thought that the people began to collect saguaro fruit four weeks before the drinking ceremony.

Lumholtz (1971, 47) observed that the saguaro harvest season lasted from the middle of June until the middle of July. A Papago informant of Nabhan (1982) said that her father used to take her to certain collecting grounds by horsedrawn wagon about sixty years ago as part of a monthlong loop through the saguaro stands around the Tucson basin. Today six weeks can be devoted to harvest by a Papago lady who demonstrates fruit gathering and processing for visiting classes and serious onlookers (Keasey 1965, 29; Fisher 1977, 2).

Thackery and Leding (1929, 409) recorded their understanding of the use of cactus camps: "The Papago Indians still have many camping places which are used year after year. . . . These camps are usually continuously occupied during the harvest season as one family after another prepares the quantity of fruit it desires and then returns home. . . . As a rule a family does not move from one camp to another."

On the other hand, Murbarger (1948, 144–48) claimed that "today mid-summer finds the tribesmen roaming over the desert camping temporarily where saguaros grow densest, gathering the harvest, moving on." The statement by Thackery and Leding would tend to indicate sequential use of one camp by several families, while the statements by Murbarger and by Nabhan's informant would indicate the reverse, sequential use of several camps by one family. Notwithstanding these possibilities, it seems likely that one family ordinarily used only one camp whenever possible.

Underhill et al. (1979, 19) discussed how families maintained separate territories for cactus picking. Such territories were not fenced like ranches or fields, but they were meant to be respected. "In the midst of the family's territory is a simple camp; an open walled sunshade, a fireplace, and little more. These, however, are somebody's place just as surely as a house is."

Herbert (1969, 3) thought that the time for the saguaro fruit harvest was "forecast by the elders according to how the season went, how much rain in the Spring and how much hot sun. Wise families go out periodically, however, to check up and see how things are coming along."

Correlation of the Harvest with the Phenology of Fruiting

The last month of the Papago-Pima year, Kai Chukalig Mashad, the month when saguaro seeds are turning black in the developing fruit, theoretically must have been determined by cutting into the immature saguaro fruits to determine the color of the seeds and, as a result, their maturity and the progression of the season.

Although there is some difference of opinion in the literature as to duration of fruit collecting, anywhere from one to six weeks, the question can undoubtedly be settled by reference to phenology of fruiting in the plant itself. The harvesting period probably coincides almost precisely with the period during which the fruit of the saguaro becomes ripe. The onset and termination of this period vary from year to year, although the duration probably changes very little. Steenbergh and Lowe (1977, 34) stated that a few individual saguaros bear their fruit as early as the last week of May or first week of June and "the number of ripe fruits increases rapidly, reaching a peak that occurs from the last week of June into the second week of July according to the year. The remaining fruits ripen and the number remaining on the plants falls abruptly during the following ten days in any year."

The Harvest Pole

The saguaro fruit harvest pole of the Papago seems generally to have been made by splicing together two long willow-like ribs of the wooden skeleton of a dead saguaro to which crosspieces were attached. Such a harvest pole is styled a *kuipaD*. Lloyd (1911, 123–24) noted that the Pima called the constellation Ursa Major by this name, "the Cactus Puller." This was also true of the Papago and even of the Seri. The same group of stars is referred to as "the Big Dipper" by Anglos, also imagining a resemblance to a domestic implement. Although most observers have reported that two ribs of saguaro comprise the *kuipaD*, Keasey (1975, 26) has encountered three-ribbed ones. Alison (1975, 4) has published a photograph documenting use of such a three-ribbed pole.

Stone (1943) recorded an instance of a harvest pole being made from "a bundle of bamboo sticks" brought from the village. Since true bamboo does not ordinarily grow in Papago country, perhaps this was

FIGURE 6.2 Fruit gatherers with *kuipaD* in hand walking out to begin picking. July 1, 1968. Arizona State Museum photograph by Helga Teiwes.

the native carrizo (*Phragmites communis*) often referred to loosely as bamboo in southern Arizona. The *kuipaD* is usually 15–24 feet long, according to Thackery and Leding (1929, 412) or up to 30 feet long according to Herbert (1955, 16). The attachment of the crosspieces has been detailed by Herbert (1969, 4): "After smoothing off the rough edges of the *kuibit* [*kuipaD*] with a knife, a *matsuguen* is made from a fifteen-inch section of tough creosote or catclaw acacia sharpened on both ends and fastened near the top of the *kuibit* at a forty-five-degree angle. The tie is made with rawhide rope, native grass or wire, whichever is handy, and the *matsuguen* hook is in place. A second *matsuguen* is secured to the midsection of the *kuibit* to pluck fruit from the shorter arms."

Thackery and Leding (1929, 309) originally stated that the crosspieces were made from creosote bush (*Larrea tridentata*) or Catclaw Acacia (*Acacia greggii*), and this has been repeated by most succeeding authors. Actual examination of museum specimens preparatory to publishing this present paper indicates that the relationship is not as specific as the literature would indicate, as other types of wood, such as saguaro itself, have apparently also been used.

The essential feature of the crosspiece seems to be that it should be tapered at each end to allow its insertion between tightly spaced fruits. The dislodging of ripe fruit seems from examination and experimentation to result from pulling or pushing and also from inserting the tapered crosspiece between tightly spaced fruit.

Setting Out to Harvest the Fruit

Once the *kuipaD* had been fashioned and the fruit was ready, the harvest was on. Herbert (1969, 3) noted that flocks of birds fed on saguaro fruit as it ripened and that sometimes a race developed between the Papago women and the birds to see which got to the fruit first. A saguaro legend of the Papago admonishes them to share the fruit with the birds and not kill them or drive them away (Enos 1945, 64–69). Keasey (1975, 27) reported that "work in the saguaro camp begins each morning at the rising of the sun, for the Indians must hurry out into the desert to get the freshly-ripened fruit before the birds and the squirrels." Setting out for fruit gathering has been described by Castetter and Underhill (1935, 20): "The women of the camp, sometimes assisted by the men, make two trips a day into an area perhaps half a mile square, which is their acknowledged territory, visiting every cactus plant in this area once in three or four days to gather the fruit as it ripens."

Once Lumholtz (1971, 77) recorded that "daily in the morning all the female members of the household could be seen proceeding on their fruitgathering expedition." Herbert (1969, 5) observed that fruit pickers "went out a good distance, then turned back toward camp and started picking in a pattern that would bring them close to camp when the baskets were full and heavy." Stone (1943, 9) described starting at the top of a hill and working down toward the camp. At the peak of the saguaro harvest a woman may walk 40 feet on the average from one cactus to another, but over 100 feet in late July as the season wanes (Nabhan 1982).

Division of Labor

Thackery and Leding (1929, 411) reported a division of labor in the cactus camp, the women collecting fruit and the men caring for the horses and keeping the camp supplied with water and firewood. Water was reportedly hauled for as far as 10 or 15 miles. Herbert (1969, 4) reported that "the young women and small children, with their *kuibits* and colorful baskets, start picking, while the men look for game and more wood and Grandmother stays at the camp to start the fires and cook." Wickham (1971, 9) stated that although a Papago man filled a bucket with saguaro fruit faster than women at the same camp, "he will not pick while I am standing close by, maybe because picking is traditionally women's work."

Harvesting the Fruit

The stance and action of a Papago woman in harvesting saguaro fruit has been reported by Wickham (1971, 7). The woman stood back from a cactus, "her hands well apart on the stick like a girl with a baseball bat, for control. She plants her feet, keeps her eyes on the fruit, and hooks with a slow rhythm. When she has cleaned all of the branches within reach, she stoops to gather them up. Her face is always protected against direct sun. She puts the best fruit in a bucket and dried hard ones into a plastic shopping bag." The dried fruits are more or less sun preserved and can be used at a later date.

Thackery and Leding (1929, 412) stated that as the fruit was dislodged and allowed to fall to the ground, no attempt was made to catch it. On the other hand, Wickham (1971, 11) noted that a Papago man could hook with one hand and catch the opened fruit with the other. Felger and Moser (1974, 261) say that Seri boys may try to catch the falling pulp in their mouths, but this seems to have been recorded as a form of play.

According to Niethammer (1974, 25), "It is important to try to catch the fruit as it falls, especially ripe fruit, for it will burst upon hitting the ground and pick up gravel that is almost impossible to clean out." When Stone (1943, 10) helped a Papago family harvest fruit, she noted that "the fruit was ripening rapidly now and must be gathered before it burst atop the plant. Then it was useless to try to rake it off."

Herbert (1969, 5) found that "two women worked as an efficient team. One pushed or pulled down the fruit pods with the *kuibit* while the other gathered them up, removed the fruit meat and put it in the basket." Niethammer (1974, 25) thought that "Saguaro gathering seems to work best in pairs: one person knocks the fruit off the cactus while the other person catches it as it falls." Thackery and Leding (1929,

412) noted that "all of the ripe fruits on a number of adjacent plants are knocked to the ground before any are picked up." Stone (1943, 9) described a Papago family knocking fruit onto a "clean canvas spread at the foot of the plant." None of the older literature mentions catching the fruit to avoid contact with the ground, and this well may be a modern hygienic concept.

Photographs have been published showing how an Anglo woman would go about harvesting fruit from the saguaro (Anonymous 1948,

FIGURE 6.3 Juanita Ahill collecting saguaro fruit west of the western unit of Saguaro National Monument. July 7, 1970. Arizona State Museum photograph by Helga Teiwes.

32) and these make an interesting commentary on differences between Papago and Anglo ways of thinking and doing things.

Filling the Basket and Returning to Camp

As the fruit is dislodged, as often as not it will split open upon hitting the ground. If not, most authors agree that the sharp calyx adhering to a fruit is used like a knife to split open fruit that remains intact. A sharp thumbnail is also used for this purpose (Thackery and Leding 1929, 412). In the old days a fingernail was allowed to grow long to be used like a knife. Niethammer (1974, 25) mentioned a small knife that can be used. Regardless of how the fruit is opened, the thumb and thumbnail are used to quickly separate the pulp and seeds from the husklike ovary wall of the fruit. Castetter and Underhill (1935, 21) stated that two motions of the thumb are used, and Thackery and Leding (1929, 410) have documented the procedure with a photograph showing a Papago woman holding a split-open fruit over a basket so that the "pulp is turned into" the basket. All authors agree that once the pulpy seed mass is in the basket, the husk is laid upon the ground with the red interior pointing toward the sky to hasten rain. According to Underhill (1951, 46), a large fruit basket was often placed in a bush while it is filled so as to keep ants out of it. Other smaller baskets may be filled and then dumped into the larger one. Lumholtz (1971, 77) wrote that two or three hours after setting out, Papago women returned to camp, each "carrying on her head a heavy harvest . . . the inside of a huge water-tight basket presented an appetizing mass of crimson fruit pulp, as well as a great amount of similarly colored juice, which would keep for a few hours only."

Herbert (1969, 5) also described return of fruit pickers to camp: "With baskets on head and *kuibit* in hand, they made the now short walk back to camp. We offered them a ride but they politely refused. Later we learned that every detail of the saguaro harvest is governed by long-standing tradition almost like a ritual, and tradition required them to bear their burdens." Thackery and Leding (1929, 412) noted that it took "most of the morning" for a Papago to gather a basket full of fruit. Castetter and Underhill (1935, 20) said that two trips per day were made. Herbert (1955, 16) reported three trips in the morning by a pair of Papago women, each carrying in a basket of fruit. Later (Herbert 1969, 5) he mentioned two trips in the morning for fruit gathering. Thackery and Leding (1929, 412) recorded collection of another basket of pulp in the afternoon "if there is need for hurry" in finishing the harvest. Keasey (1975, 27) reported on a Papago lady who would often fill "two large pails in the early morning hours, then go out again in the afternoon for more." A Papago basket full of fruit pulp weighs 15–20 pounds (Herbert 1955, 16) and holds about 14 quarts (Thackery and Leding 1929, 412).

Fruit Quantities Utilized

Thackery and Leding (1929, 411) stated that at least half of the 1,200 Papago families of that time gathered saguaro fruit and that the average family prepared three to ten gallons of syrup "and a smaller quantity of preserve" each year. A Papago lady informed Herbert (1955, 17) that "if a family has two women to gather fruit and another to keep the fires and jam-making going, the family can, during a week in camp, produce eight gallons of syrup, two gallons of jam, and a potato sack full of dried fruit and roasted seed."

Thackery and Leding (1929, 410) believed that 600 Papago families collected a total of 100,000 pounds of fruit annually. Castetter and Bell (1937, 13) reported the same number of families but 600,000 pounds of fruit, while another author (Anonymous 1948, 33) wrote that 600 families each gathered 100 pounds of fruit each summer.

If one assumes that each family had two women to collect fruit or one woman and children, or in any event that two baskets of fruit per day were collected and that each basket weighed 15–20 pounds, then three weeks of harvesting by 600 families would have yielded about 450,000 pounds of fruit pulp per year or about 750 pounds per family. Such an estimate might very well be equivalent to 600,000 pounds of fruit when the weight of the outer husk is included as well.

The Perils of Rain

The Papago do not want rain to fall during the fruit picking itself. Underhill et al. (1979, 21) pointed out that "the one complication in this period is rain which, if it comes, can wash away the ripe and opened fruit and set the picking back two or three days until a new batch has opened." Keasey (1975, 28) observed that "heavy storms knock down the ripening fruit prematurely, and the saguaro camps are not constructed to provide much shelter from bad weather."

Saguaro Fruit Preparation and Recipes

After eating the midday meal, perhaps of rabbit stew (Herbert 1969, 5) it was generally too hot to collect more fruit, so almost everyone stayed in the shade while someone, usually the grandmother, processed the fruit. Whether it arrived in a traditional basket on the head of a Papago woman or was carried in by means of a modern bucket, the fruit pulp was first cleansed of stones, sticks, dirt, or other foreign matter and immediately placed in water. Care was taken that the mixture not stand too long before cooking, lest spoilage or fermentation set in. Later, just prior to the rainmaking ceremony, fermentation will be assisted, but during fruit processing it is not wanted. Wickham (1971, 8) noted that processing had to be done immediately: "It must be started cooking before it turns sour; otherwise the syrup is inedible."

In reference to the moist pulp, Keasey (1975, 27) observed that "upon return to camp the fruit is dumped into a tub of water and is soaked for two or three hours." It is then kneaded by hand to thoroughly mix the pulp and water, and boiled one-half hour.

FIGURE 6.4 Laura Williams shaking dried saguaro fruits to remove dirt and pebbles. West of the western unit of Saguaro National Monument. July 7, 1970. Arizona State Museum photograph by Helga Teiwes.

FIGURE 6.5 Laura Williams putting dried saguaro fruit in a sack for future use. Note the basket of dried seeds, which have been separated from heated and strained fruit. Details of the cactus camp in the background. July 7, 1970. Arizona State Museum photograph by Helga Teiwes.

Use of Dried Pulp

Pulp that had already dried in the hot sun sometimes was separated from the moister pulp either when collected or as the foreign material was removed. Such dried pulp could be pressed into a mass for short-term storage. "When prepared in this manner and stored for a time, it is said by the Indians to make better flavored syrup. After being thus sun-dried the pulp is said to keep indefinitely" (Thackery and Leding 1929, 411).

Castetter and Underhill (1935, 46) did not entirely agree, noting that saguaro and other cactus fruits "could be dried and stored in jars, but not for long because they became wormy." Niethammer (1974, 24–25) found that dried saguaro fruit became wormy even when carefully dried in a screened-in porch, wrapped in aluminum foil and stored in a coffee-can with an air-tight lid.

A Refreshing Drink

Keasey (1975, 28) described how a Papago woman would offer a cup of liquid from the soaking fruit pulp to a visitor in camp. It was "thick and sweet and warm." Densmore (1921, 151) reported that saguaro syrup retained by a Papago family for household use was sometimes diluted with water to make "a pleasing drink in hot weather." Apparently, the drink could be made from dried fruit as well. In this respect, Wickham's (1971, 11) Papago informant, in reference to taking some dried saguaro fruit to "the people back home," admitted that anymore the Papago really preferred soda pop.

Preparing the Syrup

After the moist fresh pulp had soaked in water for a while and had been mashed and mixed by hand, it was boiled for one-third hour (Herbert 1969, 5), one-half hour (Keasey 1975, 27) or heated for two hours (Lumholtz 1971, 77). The soaking and heating of the pulp in water extracts the sugar from the pulp into the liquid. Sugar represents about 6.6 percent of the dry weight of a saguaro fruit and is concentrated in the fibrous material surrounding the seeds (Greene 1936, 311).

FIGURE 6.6 Mrs. Augusta Noceo processing saguaro fruit products using traditional basket and modern metal containers. Four miles southwest of Sil Nakya. July 1, 1968. Arizona State Museum photograph by Helga Teiwes.

In the old days, a special straining basket made from leaves of sotol or desert spoon (*Dasylirion wheeleri*) was used for straining the heated mixture to separate the juice from the seeds and pulp. This straining basket was placed on top of a large olla and separated from it by means of two horizontal wooden sticks. An excellent photograph documenting use of such a straining basket was published by Herbert (1955, 14; 1969, 6).

By 1929 (Thackery and Leding 1929, 412) ordinary screen wire was already being used for straining seeds and pulp from the juice. Fabric

FIGURE 6.7 Juanita Ahill mixing the saguaro fruit pulp with water before cooking. West of the western unit of Saguaro National Monument. July 7, 1970. Arizona State Museum photograph by Helga Teiwes.

of flour sacks and burlap bags also has been used for straining the heated pulp. The fabric could be attached to a frame resting on an olla, metal pot, or washtub or could be attached at four corners to sticks driven in the ground. Fisher (1977, 4) illustrated a strainer made by a Papago lady from a cloth flour sack with "sticks hemmed to the top edges of the opening to form handles."

When the juice had been removed from the pulp and seeds, the pulp and seed mixture was spread out on a canvas in the sun to dry. After cleaning the olla or metal pot of any sand or sediment (Herbert 1969, 5),

FIGURE 6.8 Mrs. Augusta Noceo adding wood to the fire while cooking saguaro pulp in a traditional olla. July 1, 1968. Arizona State Museum photograph by Helga Teiwes.

juice was returned to the fire to be cooked down to a thick syrup. The sugar concentration in the syrup was high enough to inhibit spoilage. Thackery and Leding (1929, 412) gave about an hour of cooking as the time necessary for turning the juice into syrup. Teiwes (1979) considered two to three hours more accurate. Herbert (1955, 17) reported boiling for four to five hours for thickening the juice into syrup.

Greene (1936, 309) observed the syrup to be "a deep reddish-brown in color" and similar to sorghum molasses in appearance and taste, "although it usually has a somewhat burnt taste." Thackery and Leding

FIGURE 6.9 Mrs. Augusta Noceo with ladle over olla of cooking saguaro fruit. July 1, 1968. Arizona State Museum photograph by Helga Teiwes.

(1929, 411) estimated that a gallon of saguaro syrup could be made from 20–30 pounds of fruit and noted that frequently families prepared more cactus fruit products, especially the syrup, than actually needed, "with the purpose of sale in view."

Papago saguaro syrup has been an item of commerce for many years. In 1846, Lieutenant Emory commented on the jars of "molasses expressed from the fruit" of saguaro which the Pima offered for trade (Emory 1951, 133). In 1871, Palmer (1871, 416–17) published two to five dollars as the customary price for a one-gallon jug of saguaro syrup.

Thackery and Leding (1929, 411) gave fifty cents per quart as the price in 1929. In recent years fifty cents has been the price for a sample of syrup or jam in a baby food jar. According to Nabhan (1982) the price of saguaro syrup sold by a Papago woman to a tourist has recently gone up from seven dollars to ten dollars per pint.

Preparing the Jam

Thackery and Leding (1929, 412) stated that the preserve made by the Papago was "merely the unstrained pulp cooked to the desired consistency." Herbert (1969, 6) presented a more detailed version of preparing "jam" by the Papago. Dried seedless pulp was mixed with water and added to a boiling olla of saguaro syrup. "When it swelled, forming a gelatin-like mass, it was transferred to a large mixing olla and beaten vigorously for half an hour."

Castetter and Underhill (1935, 22) gave yet another version of preparation of saguaro jam by the Papago: "All households make cactus jam, the most important sweet in their diet, and for this only a portion of the juice is drained off. The remaining moist pulp is boiled to a sweet, sticky mass, looking much like raspberry jam." Stone (1943, 10) recorded that a Papago friend added honey to saguaro jam when it was processed if a cache of the substance from wild bees could be found.

Preparing Seeds, Dehydrated Pulp, and Other Products

Once the pulp and seeds, from which the juice had been thoroughly drained, was spread out in the sun, it was allowed to dry for an entire day. Then the pulp was rubbed between the hands to separate out most of the seeds. Castetter and Underhill (1935, 22) recorded the juiceless pulp and seed mixture being "desiccated on the housetop. It is then beaten with a stick to separate the seeds."

After seeds had been removed from the pulp, the pulp was allowed to dry for another day in the sun. When the relatively seed-free pulp was thoroughly dehydrated, it could be stored for winter use or could be combined with saguaro syrup to make jam. Stored dehydrated pulp could always be boiled with water to produce more syrup at a later date.

Keasey (1975, 29) described a Papago process for further separating seeds from dried pulp when the pulp was being used for making jam. The pulp was soaked in a bucket of water and seeds floating to the surface were skimmed off with a bent piece of wire screen. "Then a short ocotillo branch is dipped in and out of the bucket, the pulp sticking to the long spines, but many of the seeds remaining behind." These seeds are then spread out to dry.

According to Castetter and Underhill (1935, ??), once seeds were dried, parched and stored, "they may be used to make meal cake or as chicken feed. One final product from the giant cactus is oil, which can be extracted from the seeds by parching them, grinding and adding water, after which a small quantity of oil comes to the surface."

Castetter and Underhill (1935, 45) reported that the Papago commonly used flour from saguaro seeds and cultivated grain to make a gruel called pinole or *waka* if simply mixed with water, and atole *orator* when boiled with salt and water. "All sorts of combinations were possible, the favorite being cornmeal with saguaro seed flour." Herbert (1955, 17) reported that roasted and ground saguaro seeds mixed with sugar made "a favorite Papago sweet." Niethammer (1974, 26) gave a recipe for a similar Papago snack using saguaro syrup in place of the sugar. Thackery and Leding (1929, 409) recorded that seeds of saguaro and organ-pipe cactus could be purchased in local markets, "but their present value is mainly for chicken feed for which they are very good."

FIGURE 6.10 Juanita Ahill straining saguaro fruit juice after the first cooking. Note the use of screen wire attached to saguaro ribs in a manner to fit perfectly over a shiny new washtub. July 1, 1970. Arizona State Museum photograph by Helga Teiwes.

Traditional Papago Saguaro Pottery

Fontana et al. (1962, 37–40, 47) presented the definitive treatment of Papago pottery associated with the saguaro. The wide-mouthed pot used for boiling the fruit was termed a *hi-to-takut*, a word which would have also been used for the smaller wide-mouth pot pictured by Herbert (1955, 14), which was used to catch the strained juice. The yet smaller narrow-necked jar that the syrup itself was stored in was

FIGURE 6.11 Two Papago ladies straining fruit pulp west of the western unit of Saguaro National Monument. July 1, 1970. Arizona State Museum photograph by Helga Teiwes.

a *si-to-ta-kut*. A larger narrow-necked jar, probably referred to by the latter term and traditionally decorated with red paint was reported by Fontana et al. (1962, 40) to be used for storing saguaro wine. This would appear to be the fermenting vessel itself since the wine is typically not stored for any time before it is drunk. Russell (1908, 124) recorded that many of the smaller decorated pots used by the Pima had been received in trade from the Kwahadk (Kohatk) and Papago, "the latter bringing them filled with cactus sirup [*sic*] to exchange for grain."

Storage Technology

According to Thackery and Leding (1929, 412), saguaro syrup to be stored for the winter "is placed in ollas and sealed tight, the seal being accomplished with a piece of broken pottery chipped into a circular form to fit the mouth of the olla. This crudely constructed cover is held in place by a preparation either of adobe mud mixed with fine grass or a transparent yellowish-brown gum of lac gathered from the 'Samo prieto' (*Coursetia glandulosa*) upon which it is produced by a scale insect (*Tachardiella fulgens*). When properly put up in this manner the syrup or preserve will keep more than a year."

Bowen (1939, 4) reported that "leaves, cornhusks, tin, broken pottery or other available material" was used for covering the top of the jar before application of the "adobe mud" air-tight seal. Pots with the air-tight seal are pictured by Herbert (1955, 17) and by Fontana et al. (1962, 38–39). Sealing ollas of jam with melted beeswax has been documented for the Papago by Stone (1943, 10). Whittemore (1893, 55) stated that saguaro jam was "stored away in small earthen jars hermetically sealed a foot or two under ground." Stone (1943, 10) recorded Papago storage of sealed jam pots by their being "swung from the rafters so nothing can get into" them.

Recipes

Several Papago recipes for saguaro fruit products have been recorded by the University of Arizona Cooperative Extension Service, by Niethammer (1974, 22–28), or can be reconstructed.

Saguaro Syrup

(Yield: 1 quart syrup)
2 buckets raw saguaro fruit pulp
1 bucket water

Clean foreign matter from fresh, ripe fruit pulp. Discard spoiled fruit. Save any sun-dried fruit for future use. Combine fruit pulp with water, squeezing the pulp by hand to break it up. Heat for 1–2 hours over a low heat or ½ hour at a boil. Remove from heat and strain to remove the juice. Clean the pot of sediment; return the juice to it for 1 hour or more of gentle boiling. Try to avoid scorching. Pour the syrup into a clean jar and store for future use.

Saguaro Seed Meal

Take drained pulp and seed (the by-products of the syrup recipe) to a sunny spot and spread on a cloth to dry for one day. Rub the pulp and seeds between the hands, allowing the seeds to fall out. Let the seeds air-dry in a shallow basket or other container for another day. Parch the seeds in a hot pan, shaking or turning well to avoid burning. Store in a clean jar. Grind the parched seed with a stone mano and metate just before using. The flour is an oily meal.

Dehydrated Saguaro Fruit Pulp

Allow the juice-free and seed-free pulp that is a by-product of making saguaro syrup and saguaro seed meal to sun dry one additional day. Store for future use as a dried fruit product.

Saguaro Fruit Jam

(Yield: 2 pints jam)

1 pint saguaro syrup

1 pint water

2 pints dehydrated fruit pulp

Soak the dehydrated pulp in the water and simmer for half an hour, stirring to obtain a uniform mixture. Add the saguaro syrup and boil until the mixture swells and thickens. Remove from heat and beat vigorously for half an hour. Seal the jam in clean jars for future use.

Atole Porridge

(Serves 4)

1 cup wheat flour

6 cups water

1 cup saguaro seed meal

1 teaspoon salt

Mix a cup of wheat flour with a cup of water. Add to 5 cups of salted boiling water. Simmer and stir until thickened. Add saguaro seed meal. Cook for 1 minute and continue to stir while cooling. Season with a little saguaro syrup or cream to taste and use like hot oatmeal or cream of wheat as a breakfast cereal.

Pinole

(Serves 1)

1 tablespoon wheat, corn, or mesquite flour

1 tablespoon saguaro seed meal

1 cup water

Stir the ingredients together and drink before the flour settles to the bottom. For a richer drink, use milk instead of water.

Papago Snacks

(Yield: 3 dozen)

5 cups saguaro seed meal

1 cup saguaro syrup

FIGURE 6.12 Juanita Ahill separating dried saguaro seeds from the fruit pulp. West of the western unit of Saguaro National Monument. July 7, 1970. Arizona State Museum photograph by Helga Teiwes.

Combine ingredients in a bowl and mix. Roll into little balls and air-dry. Sugar can be used as a substitute for most of the saguaro syrup or can be dusted onto or rolled onto the snacks.

FIGURE 6.13 Laura Williams rotating a basket to separate remaining pulp from dried saguaro seeds. July 7, 1970. Arizona State Museum photograph by Helga Teiwes.

NOTES

A Saguaro Primer

1. Hornaday 1983, 19.
2. Hornaday 1908, 72.
3. Additional competitors with the saguaro for height and mass are *Neobuxbaumia macrocephala* of Puebla, and *Pachycereus grandis* of Morelos, Puebla, and Oaxaca, Mexico.
4. The critically endangered columnar cactus *Pilosocereus robinii* (synonym *P. polygonus*) is native to the Florida Keys and may reach 9 m in height. Its branches are slender, however, and it is not an unequivocally treelike cactus.
5. Winters 2012, 424.
6. Sobarzo 1966, 212.
7. Bravo-Hollis 1978, 3.
8. See Lozoya 1984.
9. Brown 2011, 98. Felger and Moser 1985, 248, report that Seri mythology also associates saguaros with people.
10. Engelmann 1858; Engelmann 1852, 335.
11. Shreve and Wiggins 1964.
12. Britton and Rose 1919–23.

13. Bravo-Hollis 1978 (vol. 1); Bravo-Hollis and Hernando Sánchez-Mejorada 1991 (vols. 2 and 3).

Cactaceae

1. Bonfim-Patrício and Cota-Sánchez 2010.
2. Arakaki et al. 2011.
3. Valente et al. 2014.
4. Sage and Monson 1999.
5. Tapia et. al. 2017, 709 ff.
6. *Cardón* is a Spanish generic term for columnar cacti. It refers to the resemblance of their spiny fruits to *cardos*—thistles.
7. English et al. 2007; English et al. 2010a; English et al. 2010b; Bronson et al. 2011.
8. Wolf and Martínez del Río 2003.
9. Niklas and Buchman 1994.
10. Tapia et al. 2017 have referred all species of *Neobuxbaumia* and the monotypic genus *Mitrocereus* (*Pseudomitrocereus*) to the genus *Cephalocereus* in a move sure to evoke controversy. The expanded genus contains fifteen species.
11. As implied by Copetti et al. 2017.

Ecology of the Saguaro

1. For a sustained discussion of the fixing of the southern boundaries of the Sonoran Desert, see Búrquez et al. 1999, 50–60.
2. Lumholtz 1912, 46.
3. Winters 2012.
4. See especially Turner and Brown 1982, 189.
5. Brown 1982, 101.
6. Van Devender 2002, 12–16.
7. See, for example, Wallace 2002, 44, 59.
8. Unlike the common columnar cactus of the Argentine and Bolivian Andean highlands, *Echinopsis* [*Trichocereus*] *atacamensis*, which reaches elevations in excess of 3,500 m, these plants routinely experience

temperatures well below freezing but are well adapted to cold conditions. Temperatures always rise above freezing during the day.

9. Williams, Hultine, and Dettman 2014.
10. Bowers 1981, 232.
11. Nobel and Loik 1999, 151.
12. Turner, Bowers, and Burgess 1995, 146–48.
13. Turner, Bowers, and Burgess 1995, 147.
14. Kevin Hultine, in this volume.
15. Bustamante and Búrquez 2008, 1024.
16. Turner, Bowers, and Burgess 1995, 147.
17. Pierson, Turner, and Betancourt 2013, 60; Turner, Bowers, and Burgess 1995, 147.
18. Steenbergh and Lowe 1977, 107.
19. Turner, Bowers, and Burgess 1995, 305–6.
20. Yetman and Búrquez 1996, 23–31.
21. Fleming 2002, 216.
22. Medellín et al. 2018.
23. Bustamante, Casas, and Búrquez 2010; Arenas Jiménez 2015.
24. William Peachey, personal communication, 2016 19.
25. Peachey, personal communication, 2019.
26. Bustamante, Casas, and Búrquez 2010.
27. Peachey, personal communication, 2019.
28. Steenbergh and Lowe 1983, 48.
29. Steenbergh and Lowe 1983, 38; McGregor, Alcorn, and Olin 1962, 266.
30. Schmidt and Buchmann 1986.
31. Peachey, personal communication, 2019.
32. An unusually lyric and perceptive account of the longevity of saguaros is John Alcock (2009), *When the Rains Come.*
33. Drezner 2004.
34. Orum, Ferguson, and Mihail 2016.
35. Steenbergh and Lowe 1977, 50; Dan Bach, nurseryman, personal communication, 2005.
36. Félix-Burruel et al., 2019.
37. Steenbergh and Lowe 1977, 30–31; Alcorn 2009, 285–88.
38. Alcock 1990, 140–41.
39. Garvie 2003.

40. Schmidt-Nielsen 1964.
41. Bertness and Callaway 1994.
42. Franco and Nobel 1989, 870.
43. McAuliffe 1984, 319–21.
44. Ray Turner, personal communication 1995; Albuquerque et al. 2018; Alfaro-Sánchez et al. 2018.

The Anatomy and Physiology

1. Mauseth 2006; Nobel 2003.
2. Fleming, Tuttle, and Horner 1996; Wolf and Martínez del Rio 2003.
3. Barcikowski and Nobel 1984.
4. Darling 1989.
5. Gibson and Nobel 1986, 46.
6. Smith, Didden-Zopfy, and Nobel 1984.
7. Gibson and Nobel 1986, 48.
8. Nobel 1994.
9. Dubrovsky and North 2002, 48.
10. Gibson and Nobel 1986, 58.
11. Schmidt and Buchmann 1986; Wolf and Martínez del Rio 2003.
12. Mauseth and Sajeva 1992.
13. Terrazas Salgado and Mauseth 2002, 30.
14. Gibson and Nobel 1986, 6.
15. Mauseth 2000.
16. Hultine et al. 2016.
17. Williams, Hultine, and Dettman 2014.
18. Chapin, Matson, and Mooney 2002, 99–100.
19. Hall and Rao 1994, 114–15.
20. Winter and Holtum 2002.
21. Bronson et al. 2011.
22. Pimienta-Barrios et al. 2000.
23. Huber et al. 2018.

Genomics of the Saguaro

1. Sanderson et al. 2015.
2. Copetti et al. 2017.
3. Van de Peer, Mizrachi, and Marchal 2017.
4. Hernández-Hernández et al. 2011.
5. Hernández-Hernández et al. 2011.
6. Yetman 2007.
7. Hernández-Hernández et al. 2011.
8. Hernández-Hernández et al. 2011.
9. Moran 1962.
10. Bárcenas, Yesson, and Hawkins 2011.
11. Hernandez-Hernández et al. 2011.
12. Hernández-Hernández et al. 2011.
13. 1000 Genomes Project Consortium 2015.

REFERENCES

1000 Genomes Project Consortium. 2015. "A Global Reference for Human Genetic Variation." *Nature* 526 (7571): 68–74. https://doi.org/10.1038/nature15393.

Albuquerque, Fabio, Blas Benito, Miguel Ángel Macias Rodríguez, and Caitlin Gray. 2018. "Potential Changes in the Distribution of Carnegiea Gigantea Under Future Scenarios." *PeerJ* 6 (September): e5623. https://doi.org/10.7717/peerj.5623.

Alcock, John. 1990. *Sonoran Desert Summer*. Tucson: University of Arizona Press.

Alcock, John. 2009. *When the Rains Come: A Naturalist's Year in the Sonoran Desert*. Tucson: University of Arizona Press.

Alfaro-Sánchez, R., H. Nguyen, S. Klesse, A. Hudson, S. Belmecheri, N. Köse, H. F. Diaz, R. K. Monson, R. Villalba, and V. Trouet. 2018. "Climatic and Volcanic Forcing of Tropical Belt Northern Boundary over the Past 800 Years." *Nature Geoscience* 11 (12): 933–38. https://doi.org/10.1038/s41561-018-0242-1.

Alison, Kathy. 1975. "Life Among the Papago." *Arizona Highways* 51 (9): 4–10.

Anonymous. 1948. "Arizona Jam Session." *Magazine Tucson* 1: 32–33.

Arakaki, Mónica, Pascal-Antoine Christin, Reto Nyffeler, Anita Lendel, Urs Eggli, R. Matthew Ogburn, Elizabeth Spriggs, Michael J. Moore, and Erika J. Edwards. 2011. "Contemporaneous and Recent Radiations of the World's

Major Succulent Plant Lineages." *Proceedings of the National Academy of Sciences* 108 (20): 8379–84. https://doi.org/10.1073/pnas.1100628108.

Arenas Jiménez, Sebastián. 2015. *Dispersión a larga distancia vs vicarianza: variación fenotípica y genética en un cactus columnar (Stenocereus thurberi) con distribución insular, peninsular y continental en el Golfo de California.* Posgrado en Ciencias Biológicas, Universidad Nacional Autónoma de México.

Bárcenas, Rolando T., Chris Yesson, and Julie A. Hawkins. 2011. "Molecular Systematics of the Cactaceae." *Cladistics* 27 (5): 470–89. https://doi.org/10.1111/j.1096-0031.2011.00350.x.

Barcikowski, Wayne, and Park S. Nobel. 1984. "Water Relations of Cacti During Desiccation: Distribution of Water in Tissues." *Botanical Gazette* 145 (1): 110–15.

Bertness, Mark D., and Ragan Callaway. 1994. "Positive Interactions in Communities." *Trends in Ecology & Evolution* 9 (5): 191–93. https://doi.org/10.1016/0169-5347(94)90088-4.

Bomfim-Patrício, Márcia C., and J. Hugo Cota-Sánchez. 2010. "Seed Morphology, Polyploidy, and the Evolutionary History of the Epiphytic Cactus *Rhipsalis baccifera* (Cactaceae)." *Polybotánica*, no. 29, 107–29.

Bowen, Ruby. 1939. "Saguaro Harvest in Papagoland." *Desert Magazine* 2:3–5.

Bowers, Janice E. 1981. "Catastrophic Freezes in the Sonoran Desert." *Desert Plants* 2 (4): 232–36.

Bravo-Hollis, Helia. 1978. *Las Cactáceas de México.* Vol. 1. Mexico City: Universidad Nacional Autónoma de México.

Bravo-Hollis, Helia, and Hernando Sánchez-Mejorada. 1991. *Las Cactáceas de México.* Vols. 2–3. Mexico City: Universidad Nacional Autónoma de México.

Britton, Nathaniel, and Joseph Rose. 1919–23. *The Cactaceae: Descriptions and Illustrations of Plants of the Cactus Family.* Washington D.C.: Carnegie Institution of Washington.

Bronson, Dustin R., Nathan B. English, David L. Dettman, and David G. Williams. 2011. "Seasonal Photosynthetic Gas Exchange and Water-Use Efficiency in a Constitutive CAM Plant, the Giant Saguaro Cactus (*Carnegiea gigantea*)." *Oecologia* 167 (3): 861–71. https://doi.org/10.1007/s00442-011-2021-1.

Brown, Alan K., trans. and ed. 2011. *With Anza to California 1775–1776: The Journal of Pedro Font, O.F.M.* Norman: University of Oklahoma Press.

Brown, David. 1982. "Biotic Communities of the American Southwest— United States and Mexico." Special issue, *Desert Plants* 4 (1–4).

Bryan, Kirk. 1925. *The Papago Country, Arizona.* United States Geological Survey Water Supply Paper 499. Washington, D.C.: Government Printing Office.

Búrquez, Alberto, Angelina Martínez-Yrízar, Richard Felger, and David Yetman. 1999. "Vegetation and Habitat Diversity at the Southern Edge of the Sonoran Desert." In *Ecology of Sonoran Desert Plants and Plant Communities*, edited by Robert Robichaux, 36–67. Tucson: University of Arizona Press.

Bustamante, E., A. Casas, and A. Búrquez. 2010. "Geographic Variation in Reproductive Success of *Stenocereus thurberi* (Cactaceae): Effects of Pollination Timing and Pollinator Guild." *American Journal of Botany* 97 (12): 2020–30.

Bustamante, Enriquena, and Alberto Búrquez. 2008. "Effects of Plant Size and Weather on the Flowering Phenology of the Organ Pipe Cactus (*Stenocereus thurberi*)." *Annals of Botany* 102 (6): 1019–30. https://doi.org/10.1093/aob/mcn194.

Castetter, Edward, and Willis Bell. 1937. "Ethnobiological Studies in the American Southwest: IV: The Aboriginal Utilization of the Tall Cacti in the American Southwest." *University of New Mexico Bulletin* 307 (June 1): 6–21.

Castetter, Edward, and Ruth Underhill. 1935. "Ethnobiological Studies in the American Southwest: II: The Ethnobiology of the Papago Indians." *University of New Mexico Bulletin* 275 (October 15): 3–84.

Chapin, F. S., P. A. Matson, and H. A. Mooney. 2002. *Principles of Terrestrial Ecosystem Ecology.* New York: Springer-Verlag.

Copetti, D., A. Búrquez, E. Bustamante, J. L. M. Charboneau, K. L. Childs, L. E. Eguiarte, S. Lee, T. L. Liu, M. M. McMahon, N. K. Whiteman, R. A. Wing, M. F. Wojciechowski and M. J. Sanderson. 2017. "Extensive Gene Tree Discordance and Hemiplasy Shaped the Genomes of North American Columnar Cacti." *Proceedings of the National Academy of Sciences of the United States of America* 114 (45): 12003–8.

Crosswhite, Frank. 1980. "The Annual Saguaro Harvest and Crop Cycle of the Papago [Tohono O'odham] with Reference to Ecology and Symbolism." *Desert Plants* 2 (1): 2–61. Reprinted in this volume.

Darling, M. S. 1989. "Epidermis and Hypodermis of the Saguaro Cactus (*Cereus giganteus*): Anatomy and Spectral Properties." *American Journal of Botany* 76 (11): 1698–1706. https://doi.org/10.1002/j.1537-2197.1989 .tb15155.x.

Densmore, Frances. 1921. "Music of the Papago and Pawnee." *Smithsonian Miscellaneous Collections* 72 (6): 102–7.

Drezner, Taly Dawn. 2004. "Saguaro Recruitment over Their American Range: A Separation and Comparison of Summer Temperature and Rainfall." *Journal of Arid Environments* 56 (3): 509–24. https://doi.org/10.1016/ S0140-1963(03)00064-8.

Dubrovsky, J. G., and G. B. North. 2002. "Root Structure and Function." In *Cacti: Biology and Uses*, edited by Park S. Nobel, 41–56. Berkeley: University of California Press.

Emory, William. 1951. *Lieutenant Emory Reports: A reprint of Lieutenant W. H. Emory's Notes of a Military Reconnaissance*. Albuquerque: University of New Mexico Press.

Engelmann, George. 1852. "Notes on the *Cereus giganteus* of South Eastern California, and some other Californian Cactaceae." *American Journal of Science and Arts*, 2nd ser., 14 (November): 335–39, 446.

Engelmann, George. 1858. *Cactaceae of the Boundary*. St. Louis.

English, Nathan B., David L. Dettman, Darren R. Sandquist, and David G. Williams. 2007. "Past Climate Changes and Ecophysiological Responses Recorded in the Isotope Ratios of Saguaro Cactus Spines." *Oecologia* 154 (2): 247–58. https://doi.org/10.1007/s00442-007-0832-x.

English, Nathan B., David L. Dettman, Darren R. Sandquist, and David G. Williams. 2010a. "Daily to Decadal Patterns of Precipitation, Humidity, and Photosynthetic Physiology Recorded in the Spines of the Columnar Cactus, *Carnegiea Gigantea*." *Journal of Geophysical Research: Biogeosciences* 115 (G2). https://doi.org/10.1029/2009JG001008.

English, Nathan B., David L. Dettman, and David G. Williams. 2010b. "A 26-Year Stable Isotope Record of Humidity and El Niño-Enhanced Precipitation in the Spines of Saguaro Cactus, *Carnegiea Gigantea*."

Palaeogeography, Palaeoclimatology, Palaeoecology 293 (1): 108–19. https://doi.org/10.1016/j.palaeo.2010.05.005.

Enos, Susie. 1945. "Papago Legend of the Sahuaro." *Arizona Quarterly* 1:64–69.

Felger, Richard, and Mary Beck Moser. 1974. "Columnar Cacti in Seri Indian Culture." *Kiva* 39 (3–4): 257–75.

Felger, Richard, and Rebecca Moser. 1985. *People of the Desert and Sea: Ethnobotany of the Seri Indians.* Tucson: University of Arizona Press.

Félix-Burruel, R. E., E. Larios, E. Bustamante, and A. Búrquez. 2019. "Nonlinear Modeling of Saguaro Growth Rates Reveals the Importance of Temperature for Size-Dependent Growth." *American Journal of Botany* 106 (10): 1–8.

Fisher, Karen. 1977. "Papago Harvest." *Arizona Highways* 53 (6): 2–5,

Fleming, Theodore. 2002. Pollination Biology of Four Species of Sonoran Desert Columnar Cacti." In *Columnar Cacti and Their Mutualists*, edited by Theodore Fleming and Alfonso Valiente-Banuet, 207–24. Tucson: University of Arizona Press.

Fleming, Theodore, M. D. Tuttle, and M. A. Horner. 1996. "Pollination Biology and the Relative Importance of Nocturnal and Diurnal Pollinators in Three Species of Sonoran Desert Columnar Cacti." *Southwestern Natural ist*, no. 41, 257–69.

Fontana, Bernard, William Robinson, Charles Cormak, and Ernest Leavitt. 1962. *Papago Indian Pottery.* Seattle: University of Washington Press.

Franco, A. C., and P. S. Nobel. 1989. "Effect of Nurse Plants on the Microhabitat and Growth of Cacti." *Journal of Ecology* 77 (3): 870–86. https://doi.org/10.2307/2260991.

Garvie, Laurence A. J. 2003. "Decay-Induced Biomineralization of the Saguaro Cactus (*Carnegiea gigantea*)." *American Mineralogist* 88 (11–12): 1879–88. https://doi.org/10.2138/am-2003-11-1231.

Gibson, A. C., and P. S. Nobel. 1986. *The Cactus Primer.* Cambridge, Mass.: Harvard University Press.

Greene, Robert A. 1936. "The Composition and Uses of the Fruit of the Giant Cactus (*Carnegiea gigantea*) and Its Products." *Journal of Chemical Education* 13 (7): 309–12. https://doi.org/10.1021/ed013p309.

Hall, D. O., and K. K. Rao. 1994. *Photosynthesis.* Cambridge: Cambridge University Press.

Herbert, Charles. 1955. "Saguaro Harvest in the Land of the Papagos." *Desert Magazine* 18:14–17.

Herbert, Charles. 1969. "Papago Saguaro Harvest." *Arizona Highways* 45:2–7.

Hernández-Hernández, Tania, Héctor M. Hernández, J. Arturo De-Nova, Raul Puente, Luis E. Eguiarte, and Susana Magallón. 2011. "Phylogenetic Relationships and Evolution of Growth Form in Cactaceae (Caryophyllales, Eudicotyledoneae)." *American Journal of Botany* 98 (1): 44–61. https://doi.org/10.3732/ajb.1000129.

Hornaday, William. (1907) 1983. *Campfires on Desert and Lava.* Tucson: University of Arizona Press.

Huber, John, David L. Dettman, David G. Williams, and Kevin R. Hultine. 2018. "Gas Exchange Characteristics of Giant Cacti Species Varying in Stem Morphology and Life History Strategy." *American Journal of Botany* 105 (10): 1688–1702. https://doi.org/10.1002/ajb2.1166.

Hultine, Kevin R., David G. Williams, David L. Dettman, Bradley J. Butterfield, and Raul Puente-Martinez. 2016. "Stable Isotope Physiology of Stem Succulents Across a Broad Range of Volume-to-Surface Area Ratio." *Oecologia* 182 (3): 679–90. https://doi.org/10.1007/s00442-016-3690-6.

Keasey, Merritt. 1975. "Harvesting-Time in the Saguaro Forest." *Pacific Discovery* 28 (2): 25–29.

LaBarre, Weston. 1938. "Native American Beers." *American Anthropologist* 40 (2): 227–34.

Lloyd, John. 1911. *Aw-aw-tam Indian Nights.* Westfield, N.J.: Lloyd Group.

Lozoya, Xavier. 1984. *Plantas y luces en México. La Real Expedición Científica a Nueva España (1787–1803).* Madrid: Ediciones del Serbal.

Lumholtz, Carl. (1912) 1971. *New Trails in Mexico: Travels Among the Papago, Pima, and Cocopa Indians.* Reprint, Glorieta, N.Mex.: Río Grande Press.

Mahoney, Ralph. 1956. "Traditional Harvest, Giant Saguaro Cactus Yields Fruit for a Few Faithful Tribal Members." *Arizona Days and Ways*, June 24, 14–16.

Mauseth, J. D., and M. Sajeva. 1992. "Cortical Bundles in the Persistent, Photosynthetic Stems of Cacti." *Annals of Botany* 70 (4): 317–24. https://doi.org/10.1093/oxfordjournals.aob.a088480.

Mauseth, James D. 2000. "Theoretical Aspects of Surface-to-Volume Ratios and Water-Storage Capacities of Succulent Shoots." *American Journal of Botany* 87 (8): 1107–15. https://doi.org/10.2307/2656647.

Mauseth, James D. 2006. "Structure–Function Relationships in Highly Modified Shoots of Cactaceae." *Annals of Botany* 98 (5): 901–26. https://doi.org/10.1093/aob/mcl133.

McAuliffe, Joseph R. 1984. "Sahuaro-Nurse Tree Associations in the Sonoran Desert: Competitive Effects of Sahuaros." *Oecologia* 64 (3): 319–21. https://doi.org/10.1007/BF00379128.

McGregor, S. E., Stanley M. Alcorn, and George Olin. 1962. "Pollination and Pollinating Agents of the Saguaro." *Ecology* 43 (2): 259–67. https://doi.org/10.2307/1931981.

Medellin, Rodrigo A., Marina Rivero, Ana Ibarra, J. Antonio de la Torre, Tania P. Gonzalez-Terrazas, Leonora Torres-Knoop, and Marco Tschapka. 2018. "Follow Me: Foraging Distances of *Leptonycteris yerbabuenae* (Chiroptera: Phyllostomidae) in Sonora Determined by Fluorescent Powder." *Journal of Mammalogy* 99 (2): 306–11. https://doi.org/10.1093/jmammal/gyy016.

Moran, R. 1962. "*Pachycereus orcuttii*—A Puzzle Solved." *Cactus and Succulent Journal*, no. 34, 88–94.

Murbarger, Nell. 1948. "Saguaroland." *Frontiers* 12:144–48.

Nabhan, Gary. 1982. *The Desert Smells Like Rain: A Naturalist in Papago Indian Country*. San Francisco: North Point Press.

Niethammer, Carolyn. 1974. *American Indian Food and Lore*. New York: Collier Books.

Niklas, Karl J., and Stephen L. Buchman. 1994. "The Allometry of Saguaro Height." *American Journal of Botany* 81 (9): 1161–68. https://doi.org/10.2307/2445478.

Nobel, Park S. 1994. *Remarkable Agaves and Cacti*. New York: Oxford University Press.

Nobel, Park S. 2003. *Environmental Biology of Agaves and Cacti*. Cambridge, Mass.: Cambridge University Press.

Nobel, Park S., and Michael E. Loik. 1999. "Form and Function of Cacti." In *Ecology of Sonoran Desert Plants and Plant Communities*, edited by Robert Robichaux, 143–63. Tucson: University of Arizona Press.

Orum, Thomas V., Nancy Ferguson, and Jeanne D. Mihail. 2016. "Saguaro (*Carnegiea gigantea*) Mortality and Population Regeneration in the Cactus Forest of Saguaro National Park: Seventy-Five Years and Counting." *PLOS ONE* 11 (8): e0160899. https://doi.org/10.1371/journal.pone.0160899.

Palmer, Edward. 1871. "Food Products of the North American Indians." *Report of the Commissioner of Agriculture for the Year 1870,* 404–28.

Pierson, E. A., R. M. Turner, and J. L. Betancourt. 2013. "Regional Demographic Trends from Long-Term Studies of Saguaro (*Carnegiea gigantea*) across the Northern Sonoran Desert." *Journal of Arid Environments* 88 (January): 57–69. https://doi.org/10.1016/j.jaridenv.2012.08.008.

Pimienta-Barrios, Eulogio, Julia Zañudo, Enrico Yepez, Enrique Pimienta-Barrios, and Park S. Nobel. 2000. "Seasonal Variation of Net CO_2 Uptake for Cactus Pear (*Opuntia ficus-indica*) and Pitayo (*Stenocereus queretaroensis*) in a Semi-Arid Environment." *Journal of Arid Environments* 44 (1): 73–83. https://doi.org/10.1006/jare.1999.0570.

Russell, Frank. 1975. *The Pima Indians.* Tucson: University of Arizona Press.

Sage, Rowan F., and Russell Monson, eds. 1999. C_4 *Plant Biology.* San Diego, Calif.: Academic Press.

Sanderson, Michael J., Dario Copetti, Alberto Búrquez, Enriquena Bustamante, Joseph L. M. Charboneau, Luis E. Eguiarte, Sudhir Kumar, et al. 2015. "Exceptional Reduction of the Plastid Genome of Saguaro Cactus (*Carnegiea gigantea*): Loss of the Ndh Gene Suite and Inverted Repeat." *American Journal of Botany* 102 (7): 1115–27. https://doi.org/10.3732/ajb.1500184.

Saxton, Dean, and Lucille Saxton. 1969. *Dictionary: Papago and Pima to English; English to Papago and Pima.* Tucson: University of Arizona Press.

Schmidt, Justin O., and Stephen L. Buchmann. 1986. "Floral Biology of the Saguaro (*Cereus giganteus*)." *Oecologia* 69 (4): 491–98. https://doi.org/10.1007/BF00410353.

Schmidt-Nielsen, K. S. 1964. *Desert Animals: Physiological Problems of Heat and Water.* Oxford: Clarendon Press.

Shreve, Forrest, and Ira Wiggins. 1964. *Vegetation and Flora of the Sonoran Desert.* Palo Alto: Stanford University Press.

Smith, Stanley D., Brigitte Didden-Zopfy, and Park S. Nobel. 1984. "High-Temperature Responses of North American Cacti." *Ecology* 65 (2): 643–51. https://doi.org/10.2307/1941427.

Sobarzo, Horacio. 1966. *Vocabulario sonorensis.* Hermosillo: Gobierno del Estado.

Steenbergh, W. F., and C. H. Lowe. 1983. *Ecology of the Saguaro: III.* Scientific Monograph Series 17. Washington, D.C.: U.S. Department of Interior, National Park Service.

Steenbergh, Warren F., and Charles H. Lowe. 1977. *Ecology of the Saguaro: II.* National Park Service Scientific Monograph Series 8. Washington, D.C.: Superintendent of Documents.

Stone, Margaret. 1943. "Bean People of the Cactus Forest." *Desert Magazine* 6:5–10.

Tapia, Héctor J., María Luisa Bárcenas-Argüello, Teresa Terrazas, and Salvador Arias. 2017. "Phylogeny and Circumscription of *Cephalocereus* (Cactaceae) Based on Molecular and Morphological Evidence." *Systematic Botany* 42 (4): 1–15. https://doi.org/10.1600/036364417X696546.

Teiwes, Helga. 1979. Personal communication.

Terrazas Salgado, T., and J. D. Mauseth. 2002. "Shoot Anatomy and Morphology." In *Cacti: Biology and Uses*, edited by Park S. Nobel, 23–40. Berkeley: University of California Press.

Thackery, Frank, and A. R. Leding. 1929. "The Giant Cactus of Arizona: The Use of Its Fruit and Other Cactus Fruits by the Indians." *Journal of Heredity* 20 (9). 401–14. https://doi.org/10.1093/oxfordjournals.jhered.a103236.

Turner, Raymond, Janice Bowers, and Tony Burgess. 1995. *Sonoran Desert Plants: An Ecological Atlas.* Tucson: University of Arizona Press.

Turner, Raymond, and David Brown. 1982. "Sonoran Desertscrub." In "Biotic Communities of the American Southwest—United States and Mexico," edited by David E. Brown, special issue, *Desert Plants*, no. 4, 180–221.

Underhill, Ruth. 1939. *Social Organization of the Papago Indians.* New York: Columbia University Press.

Underhill, Ruth. 1946. *Papago Indian Religion.* New York: Columbia University Press.

Underhill, Ruth. 1951. *People of the Crimson Evening.* Riverside, Calif.: United States Indian Service.

Underhill, Ruth. 1979. *Papago Women.* New York: Holt, Reinhart, and Winston.

Underhill, Ruth, Donald Bahr, Baptisto Lopez, Jose Pancho, and David Lopez. 1979. *Rainhouse and Ocean: Speeches for the Papago Year.* Flagstaff: Museum of Northern Arizona.

Valente, Luis M., Adam W. Britton, Martyn P. Powell, Alexander S. T. Papadopulos, Priscilla M. Burgoyne, and Vincent Savolainen. 2014. "Correlates of Hyperdiversity in Southern African Ice Plants (Aizoaceae)." *Botanical*

Journal of the Linnean Society 174 (1): 110–29. https://doi.org/10.1111/boj
.12117.

Van de Peer, Yves, Eshchar Mizrachi, and Kathleen Marchal. 2017. "The Evolutionary Significance of Polyploidy." *Nature Reviews Genetics* 18 (7): 411–24. https://doi.org/10.1038/nrg.2017.26.

Van Devender, Thomas 2002. "Environmental History of the Sonoran Desert." In *Columnar Cacti and Their Mutualists: Evolution, Ecology, and Conservation*, edited by Theodore Fleming and Alfonso Valiente-Banuet, 3–24. Tucson: University of Arizona Press.

Waddell, Jack. 1973. "The Place of the Cactus Wine Ritual in the Papago Indian Ecosystem." *Chicago, IXth International Congress of Anthropological and Ethnological Sciences.*

Wallace, Robert S. 2002. "Phylogeny and Systematics of Columnar Cacti." In *Columnar Cacti and Their Mutualists: Evolution, Ecology, and Conservation*, edited by Theodore Fleming and Alfonso Valiente-Banuet, 42–65. Tucson: University of Arizona Press.

Whittemore, Isaac. 1893. "The Pima Indians, Their Manners and Customs." In *Among the Pimas or the Mission to the Pima and Maricopa Indians*, 52–96. Albany, N.Y.: Ladies' Union Mission School Association.

Wickham, Woodward. 1971. "Letter to Richard H. Nolte." *Institute of Current World Affairs* (September): 1–12.

Williams, David G., Kevin R. Hultine, and David L. Dettman. 2014. "Functional Trade-Offs in Succulent Stems Predict Responses to Climate Change in Columnar Cacti." *Journal of Experimental Botany* 65 (13): 3405–13. https://doi.org/10.1093/jxb/eru174.

Winter, Klaus, and Joseph A. M. Holtum. 2002. "How Closely Do the $\delta^{13}C$ Values of Crassulacean Acid Metabolism Plants Reflect the Proportion of CO_2 Fixed During Day and Night?" *Plant Physiology* 129 (4): 1843–51. https://doi.org/10.1104/pp.002915.

Winters, Harry. 2012. *'O'odham Place Names. Meanings, Origins, and Histories: Arizona and Sonora.* Tucson: Nighthorses.

Wolf, B. O., and C. Martínez del Rio. 2003. "How Important Are Columnar Cacti as Sources of Water and Nutrients for Desert Consumers? A Review." *Isotopes in Environmental and Health Studies* 39 (1): 53–67. https://doi.org/10.1080/1025601031000102198.

Wright, Harold. 1929. *Long Ago Told: Legends of the Papago Indians*. New York: Appleton.

Yetman, D. 2007. *The Great Cacti: Ethnobotany and Biogeography*. Tucson: University of Arizona Press.

Yetman, David, and Alberto Búrquez. 1996. "A Tale of Two Species: Speculation on the Introduction of *Pachycereus pringlei* in the Sierra Libre, Sonora, Mexico, by *Homo sapiens*." *Desert Plants* 12: 23–32.

INDEX

ABOUT THE AUTHORS

Alberto Búrquez is a researcher at the Instituto de Ecología, Universidad Nacional Autónoma de Mexico. His major interests focus on the study of plant animal interactions, biogeography, and the ecology of dryland plants.

Kevin Hultine received his PhD in renewable resources from the University of Arizona and has been a research ecologist with the Department of Research, Conservation, and Collections at the Desert Botanical Garden in Phoenix, Arizona, since 2011. He also holds adjunct faculty appointments in the School of Life Sciences at Arizona State University and the School of Earth Sciences and Environmental Sustainability at Northern Arizona University and is currently the plant science editor for the Oxford University Press journal *Conservation Physiology*.

Michael Sanderson is a professor in ecology and evolutionary biology at the University of Arizona. His research focuses on genomics and the evolutionary biology of plants (including cacti), with a special interest in computational methods and challenges in these areas.

David Yetman is a research social scientist at the Southwest Center of the University of Arizona. His research has focused on peoples and plants of the Sonoran Desert Region. His books include *The Great Cacti: Ethnobotany and Biogeography of Columnar Cacti* and *The Organ Pipe Cactus*. He is producer/host of the PBS television series *In the Americas with David Yetman*.

The Southwest Center Series
JOSEPH C. WILDER, EDITOR

Ignaz Pfefferkorn, *Sonora: A Description of the Province*
Carl Lumholtz, *New Trails in Mexico*
Buford Pickens, *The Missions of Northern Sonora: A 1935 Field Documentation*
Gary Paul Nabhan, editor, *Counting Sheep: Twenty Ways of Seeing Desert Bighorn*
Eileen Oktavec, *Answered Prayers: Miracles and Milagros Along the Border*
Curtis M. Hinsley and David R. Wilcox, editors, *Frank Hamilton Cushing and the Hemenway Southwestern Archaeological Expedition, 1886–1889,* volume 1: *The Southwest in the American Imagination: The Writings of Sylvester Baxter, 1881–1899*
Lawrence J. Taylor and Maeve Hickey, *The Road to Mexico*
Donna J. Guy and Thomas E. Sheridan, editors, *Contested Ground: Comparative Frontiers on the Northern and Southern Edges of the Spanish Empire*
Julian D. Hayden, *The Sierra Pinacate*
Paul S. Martin, David Yetman, Mark Fishbein, Phil Jenkins, Thomas R. Van Devender, and Rebecca K. Wilson, editors, *Gentry's Rio Mayo Plants: The Tropical Deciduous Forest and Environs of Northwest Mexico*
W. J. McGee, *Trails to Tiburón: The 1894 and 1895 Field Diaries of W J McGee,* transcribed by Hazel McFeely Fontana, annotated and with an introduction by Bernard L. Fontana
Richard Stephen Felger, *Flora of the Gran Desierto and Río Colorado of Northwestern Mexico*
Donald Bahr, editor, *O'odham Creation and Related Events: As Told to Ruth Benedict in 1927 in Prose, Oratory, and Song by the Pimas William Blackwater, Thomas Vanyiko, Clara Ahiel, William Stevens, Oliver Wellington, and Kisto*
Dan L. Fischer, *Early Southwest Ornithologists, 1528–1900*
Thomas Bowen, editor, *Backcountry Pilot: Flying Adventures with Ike Russell*
Federico José María Ronstadt, *Borderman: Memoirs of Federico José María Ronstadt, edited by Edward F. Ronstadt*
Curtis M. Hinsley and David R. Wilcox, editors, *Frank Hamilton Cushing and the Hemenway Southwestern Archaeological Expedition, 1886–1889,* volume 2: *The Lost Itinerary of Frank Hamilton Cushing*

Neil Goodwin, *Like a Brother: Grenville Goodwin's Apache Years, 1928–1939*
Katherine G. Morrissey and Kirsten Jensen, editors, *Picturing Arizona: The Photographic Record of the 1930s*
Bill Broyles and Michael Berman, *Sunshot: Peril and Wonder in the Gran Desierto*
David W. Lazaroff, Philip C. Rosen, and Charles H. Lowe, Jr., *Amphibians, Reptiles, and Their Habitats at Sabino Canyon*
David Yetman, *The Organ Pipe Cactus*
Gloria Fraser Giffords, *Sanctuaries of Earth, Stone, and Light: The Churches of Northern New Spain, 1530–1821*
David Yetman, *The Great Cacti: Ethnobotany and Biogeography*
John Messina, *Álamos, Sonora: Architecture and Urbanism in the Dry Tropics*
Laura L. Cummings, *Pachucas and Pachucos in Tucson: Situated Border Lives*
Bernard L. Fontana and Edward McCain, *A Gift of Angels: The Art of Mission San Xavier del Bac*
David A. Yetman, *The Ópatas: In Search of a Sonoran People*
Julian D. Hayden, *Field Man: The Life of a Desert Archaeologist*, edited by Bill Broyles and Diane Boyer
Bill Broyles, Gayle Harrison Hartmann, Thomas E. Sheridan, Gary Paul Nabhan, and Mary Charlotte Thurtle, *Last Water on the Devil's Highway: A Cultural and Natural History of Tinajas Altas*
Thomas E. Sheridan, *Arizona: A History, Revised Edition*
Richard S. Felger and Benjamin Theodore Wilder, *Plant Life of a Desert Archipelago: Flora of the Sonoran Islands in the Gulf of California*
David Burkhalter, *Baja California Missions: In the Footsteps of the Padres*
Guillermo Núñez Noriega, *Just Between Us: An Ethnography of Male Identity and Intimacy in Rural Communities of Northern Mexico*
Cathy Moser Marlett, *Shells on a Desert Shore: Mollusks in the Seri World*
Rebecca A. Carte, *Capturing the Landscapes of New Spain: Baltasar Obregón and the 1564 Ibarra Expedition*
Gary Paul Nabhan, editor, *Ethnobiology for the Future: Linking Cultural and Ecological Diversity*
James S. Griffith, *Saints, Statues, and Stories: A Folklorist Looks at the Religious Art of Sonora*
David Yetman, Alberto Búrquez, Kevin Hultine, and Michael Sanderson, with Frank S. Crosswhite, *The Saguaro Cactus: A Natural History*